BODIES and MACHINES

BODIES and MACHINES

MARK SELTZER

ROUTLEDGE NEW YORK AND LONDON

Published in 1992 by

Routledge
An imprint of Routledge, Chapman and Hall, Inc.
29 West 35th Street
New York, NY 10001

Published in Great Britain by

Routledge
11 New Fetter Lane
London EC4P 4EE

Printed in the United States of America on acid free paper

Library of Congress Cataloging in Publication Data

Seltzer, Mark, 1951–
Bodies and machines / Mark Seltzer.
p. cm.
Includes bibliographical references and index.
ISBN 0-415-90021-2. ISBN 0-415-90022-0 (pbk.)
1. American literature—20th century—History and criticism.
2. American literature—19th century—History and criticism.
3. Literature and technology—United States. 4. Technology and civilization. 5. Body, Human, in literature. 6. Naturalism in literature. 7. Machinery in literature. 8. Realism in literature.
I. Title.
PS228.T42S45 1992
810.9′356—dc20 92-87
CIP

British Library Cataloguing in Publication Data

Seltzer, Mark, 1951–
Bodies and machines.
I. Title
303.483

ISBN 0-415-90021-2
ISBN 0-415-90022-0 pbk

For Shirley

Contents

Acknowledgments

Work on this project was begun on a semester's grant from the American Council of Learned Societies and finished during a semester's leave at the National Humanities Center. A fellowship from the National Endowment for the Humanities, held at the Center for Literary and Cultural Studies at Harvard University, made possible much of the research. Teaching for a term at the Society for the Humanities at Cornell University, I had the benefit of an ideal group of colleagues; I am grateful to the Society and to its director, Jonathan Culler. I could not have completed this book without the generous support of these institutions.

Imagining the book would not have been possible without the support, stimulation, and challenge of the many colleagues, students, and friends who have contributed, directly and indirectly, to the project. Here I want to thank in particular: Martha Banta, Phillip Barrish, Jonathan Bishop, Fredric Bogel, Rachel Bowlby, Richard Brodhead, Laura Brown, Cynthia Chase, Walter Cohen, Jonathan Crewe, Jonathan Culler, Emory Elliott, William Flesch, Debra Fried, Michael Fried, Steven Goldsmith, Gerald Graff, Evelyn Fox Keller, Philip Lewis, Walter Michaels, Jeff Nunokawa, Andrew Ross, Neil Saccamano, Elaine Scarry, Sally Shuttleworth, Eric Sundquist, Dorothea von Muecke, and Ruth Bernard Yeazell. I want to thank also my students at Cornell, and the numerous audiences at other places, who allowed me to try out pieces of the book. My greatest debt and warmest gratitude are to Shirley Samuels, always my closest reader, dearest friend, and much more.

Parts of this book, in earlier versions, have appeared in the following journals and collections: *Sex, Politics, and Science in the Nine-*

teenth-Century Novel: Selected Papers from the English Institute, 1983–84, ed. Ruth Bernard Yeazell (Baltimore and London: The Johns Hopkins University Press, 1986) (copyright 1986 The English Institute); *New Essays on The American*, ed. Martha Banta (Cambridge: Cambridge University Press, 1987) (copyright 1987 Cambridge University Press); *diacritics*, Fall 1987 (copyright 1987 The Johns Hopkins University Press); *Engendering Men: The Question of Male Feminist Criticism*, eds. Joseph Boone and Michael Cadden (New York and London: Routledge, 1990) (copyright 1990 Routledge, Chapman, and Hall, Inc.); *American Literary History*, Fall 1991 (copyright 1991 Oxford University Press). I am grateful for the permission to reuse this material.

Introduction

Case Studies and Cultural Logistics

The Problem of the Body in Machine Culture

Nothing typifies the American sense of identity more than the love of nature (nature's nation) except perhaps the love of technology (made in America). This book traces the relays between the natural and the technological that make up what might be called the American body-machine complex.

Bodies and Machines is about the redrawing of what Thorstein Veblen called the "vague and shifting" line between "the animate and the inanimate" and between the natural and the unnatural in turn-of-the-century American culture. More exactly, it is about what the popular preacher Josiah Strong, writing in 1901, defined as "the problem of *THE BODY*" in what Veblen, among others, defined as "machine culture." *Bodies and Machines*, that is, is about how persons, bodies, and technologies are made and represented in turn-of-the-century American culture and beyond: about the "discovery" that bodies and persons are things that can be made and its implications. It traces the remaking of nature in terms of the *naturalist machine* and the remaking of individuals as *statistical persons*.

Bodies and Machines takes up, in part, the rivalry between modes of production and modes of reproduction in American machine culture: a rivalry between technological and biological modes of generating persons and things that involves at once a radical transformation in processes of production and a politics of reproduction. The book takes up, in part, the "realist" and "naturalist" fascination with the *relays* articulated between the life process and the machine process: the invention of systematic and scientific management and the work of human engineering, and the practices and discourses that manage to "coordinate" the body and the machine. It considers the discourses and practices of seeing, controlling, and managing and also the new technologies of writing and information that carry these processes.

The larger intent of this study is to resituate accounts of bodies, genders, and technologies through an historically detailed investigation of the ways in which the links between the body and the machine have focused the American cultural imagination since the later nineteenth century. *Bodies and Machines* traces how mass-produced and mass-consumed cultural forms (literary, visual, and technological) couple the body and the machine. The emphasis throughout is on the ways in which cultural forms get involved with social questions. These cultural forms include realist and naturalist writing, involving among other names, Stephen Crane and Rebecca Harding Davis, Frank Norris and Henry James, Veblen and Henry Adams, Hawthorne and George Eliot, Dreiser, Wharton, Twain, Stowe, and Zola. These cultural forms include also the mass literature of boyhood, adolescence, and the making of men: the scouting books of Ernest Thompson Seton and Baden-Powell, for instance, and the best-selling "wilding" stories of Jack London. They include visual representations such as composite photographs, scale models, and the thrilled iconography of standardized persons and things: for example, the representation of commodities in the idiom of the still life and the representation of nature in the idiom of the *nature morte*.

These are some of the practices and representations that make up the network of relays, transit points, and paths of least resistance in what has alternately been described as machine culture, as naturalist, and as the culture of consumption; and these are some of the practices and representations that bind together, as we will see, these apparently alternate descriptions in what might be called a *cultural logistics*. The social questions that these forms of writing, producing, and representing make visible include the effects of the redrawing of the uncertain and shifting line between the natural and the technological in machine culture and also the ways in which such shifts in the traffic between the natural and the technological make for the vicissitudes of agency and of individual and collective and national identity in that culture. It's this double discourse of the natural and the technological that, in short, makes up the American body-machine complex.

The interlaced essays that follow are attempts to map the psychotopography of machine culture.[1] The first part, "The Naturalist Machine," concerns the anxieties about production and reproduction, technological and biological, generated in the discourse of naturalism. Drawing on a diverse field of literary, social, and scientific discourses (including the novels of Frank Norris and the contemporary work on sexual biology and thermodynamics that informs these novels), I argue that the achievement of the naturalist novel becomes visible in the devising of a counter-model of generation that incorporates and "manages" these linked problems of production and reproduction. The second part, "Physical Capital: The Romance of the Market in Machine Culture," refers back from naturalist writing to the representations of the bodily and the economic in the earlier romance and realist writings of Harriet Beecher Stowe, Nathaniel Hawthorne, and, especially, Henry James's *The American*. It details some of the ways in which the problem of the body and of embodiment governs the "realist" account of persons and things and the intimations of the machine-likeness of persons and the personation of machines in emergent consumer culture. By way of an analysis of the notion of the "typical American"—that is, of the American *as* typical, standard, and reproducible—this part begins to draw into relation the differences and tensions between possessive individualism and market culture, on the one side, and what I call disciplinary individualism and machine culture, on the other.

These different ways of understanding the individuality of the individual focus the next part of *Bodies and Machines*. "Statistical Persons" centers on one of the dominant discourses of the nineteenth century, a discourse about agency, chance, and causation that underwrites both the emergence of nineteenth-century social science or "social physics" and styles of realist and naturalist representation: the discourse of statistics. The invention of a culture of numbers, models, and statistics and the positing of statistical persons is taken up in part by way of the chancy persons, casualties, and cases that populate the writings of Stephen Crane and in part by way of the iconography of standardization that circulates through such cases. The next part, "The Still Life," extends the examination of statistical persons to the problem of the statistical communities of mass consumption and retraces, from another angle, the relays between market culture and machine culture. Here my focus is on the problems of work and consumption (consumption in both economic and bodily senses) and on the emergence of an "aesthetics of consumption"; my examples range from the remarkable fiction of Rebecca Harding Davis to the representation of consumption, and the consumerist fascination with mobility and stillness, in the idiom of the still life. The last part of this book, "The Love-Master" examines the anthropology of boyhood and adolescence at the turn of the century and the social and cultural technologies for "the making of men." Here I am concerned with the work of systematic management, ranging from the "naturalist" turn of the scouting movements and conservation work to the extension of the "visible hand" of managerialism in the factory, school, military, and mass culture generally. The emphasis in this last part is on the writings of Jack London and Stephen Crane; the policies, for example, of Frederick Winslow Taylor and Theodore Roosevelt; and the coordination of "the natural" and the disciplines, corporal and cultural, of the machine process: that is, the operations of the naturalist machine.

This is perhaps enough to locate provisionally some of the problems that center the pages that follow. The arrangement of the parts of this study is meant to indicate at once the historical specificity of the body-machine complex in turn-of-the-century America and to pressure the appeals implicit in an overly hasty historicization of such problems. The matter of "periodizing" persons, bodies, and desires is not, I will be suggesting, separable from the anxieties and appeals of the body-machine complex but another way of negotiating the body-machine complex. What the overly hasty historicizations—the logics of equivalence and panics of reduction that continue to govern a certain style of historicism—work to disavow is the sense that *practices and discourses are never simply reducible or simply irreducible to anything else*: in brief, that "there is no equivalence without the work of making equivalent."[2] Hence the lines of force, discursive and material, I will be charting here are better understood in terms of a logistics than in terms of a logic or a general economy. That is, the cultural networks we will be mapping depend as much on the transport-work of civil and human engineering as on the abstract equivalences and "endless conversions" of the laws of thermodynamics (see Part I) or the laws of the market (see Part II). For that reason, I want in this introduction less to summarize, and therefore generalize, the accounts that follow than to set out briefly several instances that may serve to epitomize some of the terms of those accounts. The examples here are drawn from the writings of Mark Twain (*A Connecticut Yankee in King Arthur's Court* [1889]), Jack London ("The Apostate" [1906]), and Emile Zola (*La Bête humaine* [1890]).

Agencies of Expression

In every age since written language began, rhetorical forms have been to a considerable extent influenced by the writing materials and the implements which were available for man's use. This is a familiar observation in studies of the past. Is it not, then, time that somebody inquired into the effects upon the form and substance of

> our present-day language of the veritable maze of devices which have come into widely extended use in recent years, such as the typewriter, with its invitation to the dictation practice; shorthand, and, most important of all, the telegraph? Certainly these agencies of expression cannot be without their marked and significant influences upon English style.
>
> Robert Lincoln O'Brien, "Machinery and English Style" (1904)[3]

It's not hard to see that Twain's *Connecticut Yankee* is everywhere preoccupied with the conflicts, both the anxieties and the excitations, of machine culture.[4] But for the moment I am interested less in the general and explicit preoccupation with, for instance, systems of management and training than with the channels through which these conflicts themselves are directed and managed. I have in mind particularly the extraordinary series of connections made visible midway through the novel, in a chapter called "The First Newspaper." A brief setting out of the nexus of bodies, technologies, and representations that the novel perversely draws into relation here can give a preliminary sense of what Twain's version of the body-machine complex looks like and how it operates.

The chapter opens with an account of what Twain's factory boss Hank Morgan calls "the king's evil business": that is, the curing of illness through the touching of the body, or what the literalist Yankee describes merely as "fleshly contact with a king" (145). But this contact with the natural body is ostensibly something more than that: a contact, by way of the king's two bodies, between the natural body and the something beyond the natural that the body stands in for. This contact between the body and its representations is immediately translated into very different terms, terms that rewrite in the Yankee's own idiom this double discourse of the natural and its representations. The fleshly contact with the king is symbolically represented by the exchange of a piece of gold: the king's two bodies are represented by gold's, and by the commodity's, two bodies, as the representation of value and as value itself. "The piece of gold that went with the touch" is, as Morgan puts it, a matter of "skinning" and "touching" the treasury (146). And what he proposes, in this conflation of the bodily and the economic, is another version of embodiment: the substitution of minted and nickel coins for hammered and gold ones, and, more precisely, the substitution of representations and "imitation" values (the "first-rate likeness" of the king stamped on the nickel, for example) for "natural" ones (146).

Twain's brief history of the shift from feudal to market relations thus works in terms of a successive rewriting and reoccupation of a general logic of representation. It's not surprising then that the final and most provocative rewriting of the logic of representation in this series explicitly and powerfully epitomizes and radicalizes that logic: that the most explicit and most radical rewriting here makes visible the technology of writing itself. The move from the king's two bodies to the commodity's two bodies, that is, finally makes visible what might be called writing's two bodies. This writing technology involves, more exactly, what Morgan calls the "mighty birth" of the first newspaper: "I dropped my nickel out the window and got my paper" (148). The mighty birth of the newspaper is the birth of writing of "a mechanical sort . . . not written by hand, but printed" (151). But it is not merely that the introduction of the printed paper is experienced as an uncertainly material and uncertainly embodied thing ("What is this curious thing? What is it for? Is it a handkerchief?—saddle blanket?—part of a shirt? What is it made of? . . . Is it writing that appears on it, or is it only ornamentation?" [151]). The mechanically reproduced printed paper appears at the same time in the form of an unnaturally produced and unnaturally embodied person and writing's two bodies in terms of the translation of material "characters" into some more-than-material "character": "So they took it . . . and gently felt of its texture, caressed its pleasant smooth surface with lingering touch, scanned the mysterious characters with fascinated eyes" (152).

The naturalist fascination with the surfaces and skins of persons (see Part V) and the naturalist fascination with the technological making of persons (see Part I) here converge on the technology of writing and on the intimacy between technological and biological ways of making persons: "These grouped bent heads, these charmed faces, these speaking eyes—how beautiful to me! For was not this my darling, and was not all this mute wonder and interest and homage a most eloquent tribute and unforced compliment to it? I knew, then, how a mother feels when women, whether strangers or friends, take her new baby and close themselves about it with one eager impulse, and bend their heads over it in a tranced adoration that makes all the rest of the universe vanish out of their consciousness and be as if it were not, for that time" (152). Hence the bent and entranced heads looking down at the new baby, like the bent and entranced heads looking down at and scanning the printed page they are reading, correlate these rival ways of making and embodying persons, things, and representations, such that the body and writing indicate each other in circular fashion.

But the new writing-technologies instanced here pressure precisely such a fantasy of a circular or tautological relation between writing and the body. For one thing, the technology of printed writing extends what might be described as a privilege of relative disembodiment: "During all the rest of the seance," the chapter we have been looking at concludes, "my paper traveled from group to group all up and down and about that huge hall, and my happy eye was upon it always, and I sat motionless" (152). Twain thus counterposes the material circulation of the newspaper, moving from one place to the next, and the motionlessness and abstraction of Morgan's body, abstracted to the still point of the observing eye. For another, Morgan, and Twain, are everywhere fascinated by new technologies of writing, information, and communication: telegraphs, telephones, typewriters, type-setting machines. Morgan's other and natural child, in *Connecticut Yankee*, is named Hello-Central, certainly the child proper to Ma Bell; Twain not merely was famously one of the first individuals to have a home telephone and one of the first writers to work at a typewriter but also famously bankrupted himself through investments in the Paige type-setting machine. And it has frequently been noted how Twain's obsession with the Paige machine was bound up with the way in which the Paige (as opposed to the rival machine that ultimately won out) was designed to work on the model of the person, imitating the movements of a human compositor.[5]

Yet if Twain was powerfully attracted to the machine modeled on the natural person, his own way of writing—his way of writing *Connecticut Yankee*, for instance—effectively breaks the closed circuit between natural persons, bodies, and the anthropomorphisms of language. This is the how Twain described his work at the printing press on his desk, that is, his work at the typewriter:

> I am here at Twichell's house at work [on *Yankee*] with the noise of the children and an army of carpenters to help. Of course they don't help, but neither do they hinder. It's like a boiler factory for racket . . . but I never am conscious of the racket at all, and I move my feet into position of relief without knowing when I do it . . . I was so tired last night that I thought I would lie abed and rest, today; but I couldn't resist . . . I want to finish the day the [Paige] machine finishes, and a week ago the closest calculations for that indicated October 22—but experience teaches me that their calculations will miss fire as usual.[6]

What appears from one point of view as the deep self-absorption of writing (the self-absorption that makes possible the suspension of any consciousness of the scene of writing or any consciousness of the physical body that writes: "I move my feet into position of relief without knowing when I do it") appears, from another, as precisely a radical dislocation of the relays between consciousness, the body, and the act of writing (the automatisms, resistless "calculations," and misfirings of language, the type-writer's body twitching in Twichell's factory-like, nursery-like, army-like, machine-like unfinished work house).

Such moments foreground, among other things, what is experienced, at the turn of the century, as a basic difference between handwriting and typewriting. That is, such moments make visible and reciprocally intelligible a basic connection between new writing-technologies and the logistics of machine culture. As the engineer and founder of the first German typewriter business, Angelo Beyerlen, concisely stated it: "In writing by hand, the eye must constantly watch the written line and only that. It must attend to the creation of each written line . . . guide the hand through each movement. For this the written line, particularly the line being written, must be visible. By contrast, after one presses down briefly on a key, the typewriter creates in the proper position on the paper a complete letter, which not only is untouched by the writer's hand but is also located in a place entirely apart from where the hands work."[7] Hence the typewriter disarticulates the relays that allow for the circular *translation* from mind to hand to eye to mind (the translation between prelinguistic inwardness and the expressive materiality of writing, such that the eye guides what the hand does that the eye reads). As Friedrich Kittler has recently argued, in a remarkable account of the new writing-technologies at the turn of the century that make up what he calls the "discourse network" of 1900, "Underwood's invention unlinks hand, eye, and letter": "in 1900 a type of writing assumes power that does not conform to traditional writing systems but rather radicalizes the technology of writing in general."[8]

The linking of hand, eye, and letter in the act of writing by hand intimates the translation from mind to hand to eye and hence from the inward and invisible and spiritual to the outward and visible and physical, projecting in effect "the continuous transition from nature to culture." The typewriter, like the telegraph, replaces, or pressures, that fantasy of continuous transition with recalcitrantly visible and material systems of difference: with the standardized spacing of keys and letters; with the dislocation of where the hands work, where the letters strike and appear, where the eyes look, if they look at all. What these dislocations radicalize is at once a logic of standardization and a logic of prosthesis: the earliest typewriters were designed for and sometimes by the blind, as the first telephone and the first gramophone were designed by the nearly deaf (Bell and Edison). Along the same lines, Henry Ford fantasized the perfectly rationalized factory manned by the armless, the legless, and the blind (see Part V): "automatized hands work better when blind."[9]

At the close of *Connecticut Yankee*, Hank Morgan sits besieged in a cave "writing all the time." He writes, among other things, letters to his wife every day "though I couldn't do anything with the letters, of course, after I had written them" (246). The writing is, as he puts it, "almost like talking" to his wife. But the autoerotic absorption in the scene of writing ("I could sit there in the cave with my pen, and keep it up, that way, by the hour"), and the interchangeability of writing and speaking, have been reduced to a fantasy of communion. The final action of the novel is preceded by this exchange: "I tore up the paper and granted my mistimed sentimentalities a permanent rest. Then, to business. I tested the electric signals . . ." (251).

The transition in information-technologies could not be more clearly marked. But the style of transition suggests that for Twain, among others, there is nothing more sentimental than the melodramatic gesture of putting sentimentality, and the anthropomorphisms of writing, to rest for all time. It suggests also the violent immediacy promised by communication and control technologies operated by the electric signal or button: for instance, the signals and switches by which Morgan, as he expresses it, "blows up our civilization." The electric switch, ready to hand, promises to reconnect the interrupted links between conception and execution, agency and expression. Such a violent immediacy posits an identity between signal and act and an identity between communication and execution—"execution" in its several senses. It would be possible to trace out, along these lines, the fascination with the sensation of immediacy and of the pure present conveyed by the electric technology's "magical" and lightning transgression of the barriers of time and distance. Such a fascination with pure and immediate and violent action—with the pure performative that instantly connects conception, communication, and execution—is diversely expressed at the turn of the century. It is legible, for instance, in the rapid adoption of the electric chair and the "deadly current" as the socially acceptable form of legal execution in the 1880s: what Twain's friend William Dean Howells described, in 1888, as the power by which "the Governor himself might touch a little annunciation button, and dismiss a murderer to the presence of his Maker with the lightest pressure of his finger." It is legible also in the communication technology of the telephone that could order or stay that execution. (The electric chair and the telephone are scarcely separable in the cultural imaginary of "electric signals.") And it is everywhere legible in the links between Morgan's man-factories and death-factories, electric signals and body counts, and in the small magical movement of the hand that, in effect, communicates execution.[10]

Bodies in Motion

> . . . the pure Yankee, in a word, is not only a worker, he is a migratory worker. He has no root in the soil . . . he is always in the mood to move on, always ready to start in the first steamer that comes along from the place where he has just now landed. He is devoured with a passion for movement; he must go and come, he must stretch his limbs and keep his muscles in play. When his feet are not in motion, his fingers must be in action; he must be whittling a piece of wood, cutting the back of his chair, or notching the edge of the table, or his jaws must be at work grinding tobacco. Whether it be that continual competition has given him the habit, or that he has an exaggerated estimate of the value of time, or that the unsettled state of everything around him keeps his nervous system in a state of perpetual agitation, or that he has come so from the hands of nature—he always has something to do, he is always in a terrible hurry. He is fit for all sorts of work except those which require a careful slowness. Those fill him with horror; it is his idea of hell. "We are born in haste," says an American writer . . . "Our body is a locomotive going at the rate of twenty-five miles an hour; our soul, a high pressure engine . . ."
>
> Michel Chevalier,
> *Society, Manners, and Politics in the United States* (1839)[11]

The turn-of-the-century fascination with technologies of writing and representation inheres not simply in the notion that machines *replace* bodies and persons (as in, for example, Melville's earlier account of the replacement of biological by mechanical production and of bodies by paper, in "The Tartarus of Maids" story about the Berkshire paper mills [1855]); nor is it accounted for primarily in the notion that persons are *already* machines (although this is what centers, for example, such cases as Villiers's fantasies of Edison, the new woman, and other artificial or prosthetic persons, in *L'Eve future* [1886]); nor even is it "covered" in the notion that technologies *make* bodies and persons (as in, for example, Frederick Winslow Taylor's or Baden-Powell's programs for the systematic making of men).[12] Opposed on part of their surface, what makes it possible for these powerfully insistent, but not entirely compatible, notions to communicate on another level is the radical and intimate *coupling* of bodies and machines.[13] Or, as Jack London represents the "perfect worker" and "perfect machine" who is the central figure in his short story, "The Apostate": "There had never been a time when he had not been in intimate relationship with machines."[14]

"The Apostate," which was originally subtitled "A Parable of Child Labor," tells the story of a young factory worker—the perfect worker and perfect machine—who is driven by the speed and repetitiveness of his machine-work into what, by the late nineteenth century, had come to be called "pathological fatigue" or "the maladies of energy."[15] From the start he has been intimate with machines: from the moment of his birth, in the mechanical womb of the "loom room," his mother "stretched . . . out on the floor in the midst of the shrieking machines" (801) to his achievement of a "machine-like perfection," such that, in his Taylorized body, "all waste movements were eliminated. Every motion in his thin arms, every movement of a muscle in the thin fingers, was swift and accurate" (807). But the result of his working at "high tension" was that "he grew nervous" and the result of this case of American nervousness is his complete turning from work and turning apostate to the work ethic. Yet this version of hysteria or "neurasthenia is a kind of inverted work ethic": as Anson Rabinbach has recently shown in his history of "the human motor," neurasthenia is an "an ethic of resistance to work or activity in all its forms."[16] London's apostate is one case study in the neurasthenic inversion of the work ethic (and it would be possible to reveal the lines of connection in case studies ranging from De Quincey to Melville's Bartleby, from London's Apostate to Proust). London's worker at last assumes a position of absolute "movelessness" (one of London's complex words). Or, rather, since his body and nerves keep working "automatically" and apart from his own volition ("Sometimes he dozed, with muscles that twitched in his sleep"), he utterly severs movement and volition ("When awake, he lay without movement" [816]). Which is to say that the apostate to the work ethic in effect reinvents, in inverted form, the principle of movement without volition, which is here the very principle of machine-work, as the pathologized and moveless state of his resistance (see Parts IV and V).

London's case study in fatigue thus epitomizes the transposition of the character of the energy-converting machine and the character of the natural body: not the demotion of the living body to the machine but their intimate correlation.[17] Studies in the maladies of energy at every point coordinated bodies and technologies, the life process and the machine process. As the *Yearbook* of the Smithsonian Institution in 1911 noted (and as if such "coincidences" continued to surprise),

"The curves of fatigue for metals coincided in a remarkable way with the curves of fatigue for muscular effort."[18] But I am, for now, less interested in London's story as a parable of the work ethic than I am in a somewhat different parable the story traces. "The Apostate" insistently focuses in on the representation and calculation of the motions of bodies and the motions of machines. It brings into focus, more exactly, the relations thus drawn between representing and calculating. This "coincidence" of representation and calculation and its implications (a coincidence to which I will return in the chapters that follow) is what I want provisionally to set out here.[19]

It is not merely that London's writing, and naturalist writing generally, everywhere notes numbers and intervals of time; calibrates time and motion; measures and decomposes values, distances, and actions into intervals, sequences, and statistics. The life process and the work process are decomposed into body counts, repetitive motions, and abstract quantities (to take one example from this story—but it is the multiplication of instances that italicizes moments such as these: "There had been several great events in his life. One of these had been when his mother bought some California prunes. Two others had been the two times when she cooked custard. These had been events" [808]). It is, however, not simply that such moments in effect reduce events to their material or physical components: reduce the representation of persons or actions to a physics or psychophysics. What from one point of view appears as the reduction of persons and actions to sheer physicality or materiality, appears, from another, as the abstraction of bodies, individuals, and "the natural" itself. On the one side, that is, we find the insistence on the *materiality* or *physicality* of persons, representations, and actions in naturalist discourse; on the other, the insistent *abstraction* of persons, bodies, and motions to models, numbers, maps, charts, and diagrammatic representations. This is the double discourse—the dematerialized materialism or what Stephen Crane calls the "transcendental realism"—that, as we will see, at once represents and produces the composite or statistical persons, the working models and living diagrams, and the unnatural Nature of naturalism.

This is also what makes it possible for *production* and *representation* to indicate each other in machine culture. That is, naturalist writing instances a fundamentally different understanding of the work process and of the relation of writing and representing to the work process. What this involves in part is the incorporation of the representation of the work process into the work process itself. But, beyond that, it involves the incorporation of the representation of the work process *as* the work process itself (see Part V).[20] Consider, for example, the ways in which representing and producing (picturing and making) indicate each other in London's "The Apostate": "On the surface of the cloth stream that poured past him, he pictured radiant futures wherein he performed prodigies of toil, invented miraculous machines . . . in the smiling future his imagination had wrought into the steaming cloth stream" (808). There is something more operating here than the reduction of mind to the matter of work (the reduction of the stream of consciousness to the cloth stream, for example), although this is certainly where London's sentiments about the degradation of labor are located. The materialist reduction is signaled in London by the becoming-visible of the materiality of writing (eg. "steam"/"stream"). But what also becomes legible here is the identification of technologies of writing, the material streams of information and representation, and "flow" technologies of production: the conflation, for instance, of the cloth stream and the continuous sheets of paper at which London worked.

What this suggests is the radical entanglement of relations of meaning and relations of force, and, along the same lines, the radical entanglement of writing, bodies, and mechanics. Thus London, in his autobiographical *John Barleycorn*, represents physical work in these terms: "sitting at my machine, in the stifling, shut-in air, repeating, endlessly repeating, at top speed, my series of mechanical motions."[21] Thus London represents writing as working at the machine and the machine-work of writing in terms of the violent and repeated motions and impressions made by the body itself when, as London puts it, "the old flesh-machine is running":

> And then there was the matter of typewriting. That machine was a wonder . . . I'll swear that machine never did the same thing the same way twice. Again and again it demonstrated that unlike actions produce like results . . . The keys of that machine had to be hit so hard that to one outside the house it sounded like distant thunder or some one breaking up the furniture. I had to hit the keys so hard that I strained my first fingers to the elbows, while the ends of my fingers were blisters burst and blistered again. Had it been my machine I'd have operated it with a carpenter's hammer . . . The worst of it was that I was actually typing manuscripts at the same time I was trying to master that machine. (1049)

The "matter" of typewriting decomposes writing to its physics of force and motion and decomposes actions such that force and volition are unlinked ("unlike actions produce like results"). But this is not simply to reduce writing to matter; rather, it is to imagine matter writing itself.

That is, the typographic method of writing registers and translates what the founder of time/motion studies Etienne-Jules Marey called the "natural language of the phenomena themselves, so much superior to all other modes of expression."[22] What the graphic method of measuring the dimensions of bodily motions achieves, and what the naturalist "mechanics of fiction" relays, is the natural iconography of bodies in motion. These technologies of registration or mechanical inscription amount to a sort of automatic writing that "united the body's own signs (pulse, heart rate, gait, the flapping of wings) with a language of technical representation."[23] What this discloses, in Marey's terms, are the mechanical inscriptions written in "*une langue inconnue*" of the body itself, the unknown language written and deciphered by new graphic technologies (Marey's *la méthode graphique*). "When the eye can longer see, the ear cannot hear, or touch cannot feel, or even when the senses appear to deceive us, these instruments perform like a new sense with astonishing precision": what the writer's signs here trace is thus not the writer's "own" desires or the writing self but the prosthetic language of the body and the mechanical and automatic inscription of the body's forces.[24]

London notoriously was a writer who counted his words (he produced a thousand words a day) and measured his output in sheer numbers. As the literary historian Ronald E. Martin observes: "Significantly, when [London] talked about his writing he was preoccupied with the quantitative rather than the qualitative aspects of his craft. How many words a day he was able to write, how many copies his books sold, how much money they made for him—these were his main concerns; rarely did he discuss quality, rarely differentiate between his works on that basis."[25] But it should by now be clear that such a preoccupation with quantities is something other than a sacrifice of quality for quantity: it registers the decomposition of qualities into quantities and the positing of, and experience of, an equivalence between writing and the machine process. When London's apostate stops working he starts writing, but working at the machine and writing both appear as the "figurin' " of moves and calculations: "Sometimes his lips moved. He seemed lost in endless calculations . . . Next morning, after the day grew warm, he took his seat on the stoop. He had pencil and paper this time with which to continue his calculations, and he calculated painfully and amazingly . . . Moves. I've been movin' ever since I was born . . . That makes twenty-five million moves a year, an' it seems to me I've been movin' that way 'most a million years" (812–14). This is, sketched in very preliminary form, the psychophysics of bodies in motion that makes up the naturalist mechanics of writing.

Intimacy with Machines

> Does there exist, anywhere on this earth, a being conceived in the joys of fornication and born in the throes of motherhood who is more dazzling, and more outstandingly beautiful than the two locomotives recently put into service on the Northern Railroad?
>
> J.K. Huysmans, *A Rebours* (1880)

> The man, for the time being, becomes a part of the machine in which he has placed himself, being jarred by the self-same movement, and receiving impressions upon nerves of skin and muscle which are none the less real because they are unconsciously inflicted.
>
> *The Book of Health* (1884)

When London's apostate turns from both work and writing, he walks down "a leafy lane beside the railroad track" and then this "nameless piece of life . . . pulled open the side-door of an empty box-car and awkwardly and laboriously climbed in" (815–16). The juxtaposition of the leafy lane and the railroad track concisely counterposes the natural and the technological and the garden and the machine.[26] But what makes the understanding of such a counter-position as opposition misleading, we have seen, is precisely the intimacy between the natural and the technological in naturalist or machine culture. Moreover, if machine-work makes the apostate turn neurasthenic or hysteric, his "escape" by entering into the compartment of the train seems less an alternative to either machine-work or neurasthenia than another way, like the way of the hysteric, of experiencing one's body (or "piece of life") set in motion apart from one's own intentions. There is a striking resemblance between the photographic record of hysterics in Charcot's *Iconographie photographique de la Salpêtrière* (1878–1881) and Eadweard Muybridge's photographs of bodies in motion in his *Animal Locomotion* (1883).[27] The cross-influences among Charcot's studies in hysteria, Muybridge's gridding of moving bodies, and the graphic time-motion studies of Marey and, somewhat later, Frank Gilbreth, make visible the unlinkings of motion and volition that allow hysteria, locomotion, and machine-work to communicate with each other. The principle of locomotion which in liberal market culture is the sign of agency is in machine culture the sign of automatism (see Part IV). Or rather, since the recalcitrant tensions between the imperatives of market culture and machine culture are central here (see Parts II and IV), what must be considered is how the uncertain status of the principle of locomotion precipitates the melodramas of uncertain agency and also what amounts to an erotics of uncertain agency. Not surprisingly, the crisis of agency and its appeals are most evident in the figure of the railway locomotive itself.

The steam locomotive couples what Chevalier called the American "passion for movement" and the mechanization of motive power; and the relays between passion and mechanism are evident in the pervasive sexualization of steam technologies and trains in nineteenth-century discourse (see Part I). On another level, the railway system combines mobility and incarceration, confining still or stilled bodies in moving machines directed by mechanical prime movers. The railway, like the elevator, or like (in its recreational form) the Ferris wheel, puts stilled bodies in motion. What these mobile technologies make possible, in different forms, are the thrill and panic of agency at once extended and suspended. More precisely, the railway system is, as Michel de Certeau observes, "organized by the gridwork of technocratic discipline, a mute rationalization of laissez-faire individualism."[28] Combining and adjusting the disciplines of organized movements, classes, time-tables, and classifications, on the one side, and the market principle of competitive agency in mobility, on the other, the railway journey and the railway system powerfully epitomize what I will be examining in terms of the differences and rivalries between styles of competitive and disciplinary individualism. The railway epitomizes also the tensions between the bodily and the visual that will concern me in what follows. The transport of still persons across landscapes converts landscapes into the arrested and unrooted nature of the still life. That is, the panoramas in perspective, seen through the glass frame of the window, reduce motion to the shift of the gaze, and the shift of the gaze reduces landscape to "scenery" and to the cinematographic illusionism of the *trompe l'oeil* (see Part IV).[29]

There is perhaps no more powerful registration of *une langue inconnue* of the railway system, its logics and its erotics, than Zola's novel *La Bête humaine*. Zola's novel tracks not merely the understanding of bodies as thermodynamic mechanisms (from the start, the body as a steam engine and heat exchange system) and not merely the understanding of persons in terms of "human-machines systems"—what Zola calls "human dolls" and "metal beings."[30] *La Bête humaine* maps the relays and exchanges between control-technologies of transportation and communication (railways and telegraphs), control-bureaucracies of information and policing (the judiciary), and, above all, the crossing-points between these technologies and the violent erotics these crossings generate. If it seems that everyone in the novel works for the railway, has a connection with the legal system, and has recently planned or committed a sexual murder, what this indicates are the violent desires incited by the systemic, the repetitive, and the automatistic: desires constituted by the obsessional and the machinic, by "the tinklings of the apparatus" and by "obsessions [that] ticked on like a clock" (318). The novel distributes anxieties about agency, and the desire for possession and self-possession, across physical landscapes. The places, psychological and geographical, where "paths have crossed" (319) are the crossing paths of the natural and the technological, the crossings between interior states and external systems, and between bodies and persons and machines. These crossings map but also *realize*, make visible and material, what I have been calling the psychotopography of machine culture.

Zola's novel, that is, traces the straight lines drawn across and cut across natural landscapes by machine-technologies (for example, "the narrow garden cut in two by the railway" [311]). What motivates desire in the novel is the imperative "to go in a straight line" (69), to proceed "further and further" in a direct line, always "gazing at the line" (286). (This is a desire not limited to, but most explicit in, Jacques Lantier, *mécanicien de locomotive.*) These straight lines drawn and cut across abstracted and map-like places correlate technologies of writing and the machine process, correlate representation and production, models and practices, word counts and body counts. These are the correlations and relays that will be examined and specified in the case studies that follow. The railway system's straight lines have affinities with, for example, the steamship "liner" that, for Frank Norris, represents the "mechanics of fiction" (see Part I). This, in turn, anticipates the "toil of trace and trail" that traces lines across the abstract white spaces of Jack London's great white male North (see Part V). And this, albeit with markedly different implications, binds together Stephen Crane's obsession with regimented lines of bodies, the torn bodies of the corps moving in a "moving box" (see Part III) and London's obsession with the body-wounding and machine-like movement of lines of print across the typewritten page.

The naturalist mechanics of writing is thus not simply a reduction of action to "the scene of writing" (that is, to the scene of the writer's self-absorption or self-reference); nor is the naturalist mechanics of writing simply a transparency through which appears the materiality of writing in general, any more than Norris's or Crane's or London's or Zola's straight lines and liners are transparent "figures" of writing. The foregrounding of the scene of writing and the materiality of writing in these cases indicates how writing and mechanics in naturalist discourse refer back to each other in circular fashion; it registers the fascinated, and at times excruciated, coupling of the work of writing and the workings of the body-machine complex (see Part III). Hence Zola's *La Bête humaine* holds steadily in view and in taut relation the lethal "instrument" of writing and the "apparatus" of desire, transfixed by the erotics of the bruised body that painfully and mechanically writes: "Instrument of love, instrument of death. 'Write, write.' She wrote painfully, with her poor hurt hand" (44). ["*Instrument d'amour, instrument de mort. 'Écris, écris.' Et elle écrivit, de sa pauvre main douloureuse, péniblement.*"][31]

Finally, the desire to go in a straight line couples the logics and the erotics of machine culture. In Zola's novel, erotic desire takes the form of a "destruction for fuller possession" (338). And fuller possession takes the form of the desire to make intention and act, cause and effect, line up exactly: "The two murders were coupled together. Was not one the logical outcome of the other" (332) ["*et les deux muertres s'étaient rejoints, l'un n'était-il la logique de l'autre*"]. By this logic, "the past was connected to the present" such that "the first . . . led mathematically to the second" (345). More exactly, the coupling of logicism and eroticism is localized in a male hysteria that moves to "fix" agency through a violence directed at the female body: "the knife nailed the question in her throat" (331). But such a fixing of agency at the same time appears as the radical suspension of agency: the relocation of agency in the repetitive compulsions of "the inexorable laws of murder" and in the automatistic "desire of his hand which was having its own way" (331). The historical desire to make the past connect with and lead to the present thus crosses paths with the hysterical desire to "nail" agency. By the close of the novel, the historical and the hysterical are scarcely distinguishable. The "two cases" of sexual murder are explicated in this brief history of the close of the Second Empire: "Since the noisy success of the plebiscite the country had been in a continual state of hysteria, like the unstable condition which precedes any catastrophe. Society, in the closing phase of the Empire, was pervaded in politics and above all in the press by a continual restlessness, a highly-strung condition in which even joy took on an unhealthy violence" (349). The crossing paths of pleasure and violence, of interior and external states, of eroticism and logicism in these cases: these crossings intimate the appeals and violences induced both in an overly hasty historicization and in an overly hasty hystericization, and in the attempt to make one simply the logical outcome of the other.[32]

"We should not hurry to divide 'nature' from 'culture,' " the sociologist of science Bruno Latour has recently observed: "Scallops also find that nature is a harsh taskmaster—hostile, nourishing, profligate—because fish, fishermen, and the rocks to which they attach themselves have ends that differ from those of scallops."[33] Such an attribution of "ends" across the lines that divide the natural and the cultural, the organic and the inorganic, violates at the least the principle of scarcity with respect to agency and personhood that the nature/culture division serves to protect. Such a promiscuousness with respect to agency and personhood takes the form of what might be called a miscegenation of the natural and the cultural: the erosion of the boundaries that divide persons and things, labor and nature, what counts as an agent and what doesn't. And such an erosion of boundaries immediately evokes the melodramas of uncertain agency everywhere rehearsed in the work of dividing "nature" from "culture." These are some of the lines of force that I will be following out in this study of the problem of the body in machine culture, its logics and its erotics.

Part I

The Naturalist Machine

In these pages I will be concerned with a cluster of anxieties, at once sexual, economic, and aesthetic, that seems to be generated in the late nineteenth-century "naturalist" novel. More specifically, I will be considering an insistent anxiety about production and generation—generation of lives, powers, and representations—that marks this fiction. If, as the industrialist Cedarquist observes in Frank Norris's *The Octopus*, "the great word of this nineteenth century has been Production," production, both mechanical and biological, also troubles the naturalist novel at every point.[1] I want to suggest that the achievement of the naturalist novel appears at least in part in the devising of a counter-model of generation that incorporates and works to manage these linked, although not at all equivalent, problems of production and reproduction. The counter-model is what might be called the naturalist machine. Although I would argue that such a model operates in a wide range of naturalist texts, I want for the moment to take most of my examples from the work of the American novelist who most conspicuously and compulsively displays both these anxieties about generation and the aesthetic machine designed to manage them—Frank Norris.

Production and Generation

Near the very end of *The Octopus*, the railroad agent and speculator S. Berman is buried alive beneath a "living stream" of wheat, "this dreadful substance that was neither solid nor fluid" (2:353). The logic of such a fate is clear enough; it is explicitly represented as a productive or hyperproductive nature's revenge on what Norris throughout presents as the unnatural and nonproductive abuses of the speculator. If Berman is bodily, as the speculator Curtis Jadwin in Norris's *The Pit* is financially, buried by a flood of wheat, this is a sort of turn of nature on the speculator who attempts to appropriate or corner her living produce. But such a homeopathic logic fails to account for a somewhat different sort of uneasiness that "this dreadful substance" presents.

Just before this "inevitable" (2:355) end of the speculator, Berman discusses the terms of his deal in wheat with one of his agents. "This deal is peculiar," he observes, "it's a queer, mixed up deal." What is "queer" about the deal, it would seem, is its direct, unmediated character. "I'm not selling to any middleman," Berman explains: "I've got to have some hand in shipping this stuff myself" (2:328). But this exclusion of—or what the rancher Magnus Derrick earlier calls "stranding" of—the middleman turns out to be a little more complicated. Berman claims that he is "acting direct," but in fact, as he goes on to say, he's "acting direct with these women people," with a "lot of women people up in the city" who have contracted for the wheat as part of a project to relieve a famine in Asia (2:327).[2] What relation do the middlemen, on the one side, and the "women people," on the other, have to the logistics of production in the novel?

The Octopus, as I have indicated, invokes a traditional "agrarianist" opposition of producer and speculator or middleman, but such an opposition is not consistent with the account of production that the novel ultimately supports.[3] Not surprisingly, this account is articulated by the railroad titan Shelgrim, the novel's middleman par excellence. Shelgrim's reply to Presley's protest against the railroad's usurpation of the ranches and their wheat is in part a self-exonerating appeal to an invisible hand guiding the economy: "Where there is a demand sooner or later there will be a supply. . . . Blame conditions not men" (2:285). The ideological character of such a defense is evident. But Shelgrim's defense goes a bit further, in a manner that both more effectively threatens the opposition between producer and speculator that grounds Presley's protest, and concisely enunciates what Norris ends by endorsing as the "larger view" (2:361). "Try to believe this—to begin with—*that railroads build themselves*. . . . Mr. Derrick, does he grow his wheat? The Wheat grows itself. What does he count for? Does he supply the force?" (2:285). Crucially, by the logic of this larger view, there are no producers at all: all are middlemen, equally subject to or carriers of uncontrollable forces. One consequence of such a view is, of course, a radical emptying of the category of production—the very category that the social-economic "protest" the novel might be seen to embody centrally requires.

Yet if production does not have a secure place in Norris's economic theory, the novel does offer an account of the "enigma of growth" and "mystery of creation" (2:343). *The Octopus* offers in fact *two* competing accounts of the force of production and generation. On the one side, and not unexpectedly, there is the mother, at once the "the crown of motherhood . . . the beauty of the perfect woman" and the mother-land, "palpitating with the desire of reproduction" in the "hidden tumult of its womb" (2:215, 1:122), "the great earth, the mother, after its period of reproduction, its pains of labour, delivered of the fruit of its loins, [sleeping] the sleep of exhaustion in the infinite repose of the colossus" (1:123, 2:342–43). On the other side, Norris represents the technology of the steam machine, "the jarring, jolting, trembling machine . . . [t]he heroic embrace of a multitude of iron hands, gripping deep into the brown, warm flesh of the land" (1:124–25). These twinned representations of generative power—the machine and the mother—are suspended in relation to each other in the novel, as linked but competing principles of creation.

Norris's representation of steam power as generative, in *The Octopus* and throughout his work, participates in a more general field of middle and later nineteenth-century discourses on the procreative force of the machine. As Perry Miller has argued, in the American representation of steam power in this period—"the pure white jet that fecundates America"—the "imagery frequently becomes, probably unconsciously, sexual, and so betrays how in this mechanistic orgasm modern America was conceived."[4] Such an association of steam power and generation—what writers of the midcentury described as "the marriage of water and heat" that inseminates the "body of the continent"—is part of a larger celebration of technology by which Americans viewed the machine, and especially the steam engine and dynamo or "generator," as a "replacement for the human body."[5] More precisely, this "replacement" registers also a displaced competition between rival sexual forces, between what Norris, for instance, calls the "two world-forces, the elemental Male and Female" (1:125).

The nineteenth-century obstetrician and gynecologist Augustus Kinsley Gardner opens his lecture *The History of the Art of Midwifery* (1852) by strangely coupling these two rival principles of generation:

> From the foundation of the world man has been born of women; and notwithstanding that his inventive genius has discovered steam, the great Briareus of the nineteenth century, and harnessed him to his chariot, and sends lightning to do his bidding over the almost boundless extent of the world, yet we cannot hope that any change may be affected in this particular.[6]

Despite or "notwithstanding" the rhetorical disclaimers, it is not hard to see that Gardner's purpose, here and elsewhere, is to "replace" female generative power with an alternative practice, at once technological and male. As Graham Barker-Benfield argues in his study of attitudes toward women and sexuality in Victorian America, one of the governing impulses promoting the medicalization of women and childbirth in the nineteenth century was the desire "to take charge of the procreative function in all of its aspects."[7] And one of the markers in this takeover was an attack on female midwifery and the substitution of the professional and male technology of obstetrics. This amounted to, in effect, not merely an increasing male management of procreation but also a revaluing of the midwife's position in relation to reproduction. That is, if the obstetrician replaces the (female) midwife, his confiscation of the generative function places him as the governing middleman of reproduction, and governing precisely because of his position as middleman. And if Norris's image of the earth-mother is centrally obstetrical—the mother "delivered," after the pains of labor—this delivery, by the iron "knives" of the steam harvester, represents, I want to suggest, a collateral reassertion of the middleman—again, the position of all men in Norris's account—in the process of production.

It is tempting to read this revaluation of the middle*man* as a compensatory male response to a threatening female productivity, and, as Barker-Benfield, among others, has suggested, this is certainly a significant part of the story. But the notion of, and promotion of, a rivalry between "male" and "female" forces, and the consequent underwriting of what appears as an absolute differentiation of gender powers and "principles," may in fact function as ways of managing the anxieties about production we have been sketching. Put another way, we must consider the ways in which difference itself may be produced and deployed as a strategy of control and as part of a more general economy of bodies and powers. One of the narrative tactics that supports this economy is the displacement or rewriting of generative power, and one of the supports of such tactics is the resolutely abstract account of "force" that governs the naturalist text.

The passage on the earth-mother's delivery is, in characteristic Norris fashion, repeated several times in the novel. The final repetition is quickly succeeded by two remarkable rewritings of the force of (re)production. The mother initially appears to extradite men from a place in what Norris calls "the explanation of existence": "men were nothings, mere animalcules." The negation of male power is evident.[8] But through the double operation of negation and reinvestment that I have already noted in Norris's explanations of production, a curious revaluation follows. If the titanic mother reduces men to "nothings," "for one second Presley could go one step further":

> Men were naught, death was naught, life was naught; FORCE only existed—FORCE that brought men into the world, FORCE that crowded them out of it to make way for the succeeding generation, FORCE that made the wheat grow, FORCE that garnered it from the soil to give place to the succeeding crop. (2:343)

The colossal mother is thus rewritten as a machine of force that brings men into the world, "the symphony of reproduction" as "the colossal pendulum of an almighty machine." And crucially, if the mother is merely a "carrier" of force, the mother herself is merely a medium—mid-wife and middleman—of the force of generation.

Such a capitalizing on force as a counter to female generativity in particular and to anxieties about generation and production in general may help to explain, at least in part, the appeals to highly abstract conceptions of force in the emphatically "male" genre of naturalism.[9] More specifically, and despite the instabilities and contradictions in the naturalist conception of force, it does not take much interpretive pressure to see that this discourse of force, crucial to the naturalist style of power, is essentially the discourse of thermodynamics.[10] I am referring not merely to the centrality of steam power in the naturalist text (preeminently Zola's) but also to the "laws" that govern what I have called the naturalist machine. The two fundamental principles of thermodynamics—the law of conservation and the law of dissipation—operate, I want to argue, both thematically and formally in the naturalist narrative. The pertinence of the second law—positing the irreversible degradation of usable energy in any system and hence an inevitable systemic degeneration—to the naturalist doctrine of degeneration is immediately evident, and I will be discussing this "application" of the second law in the section following. But here I am more concerned with some surprising consequences of the first law for the naturalist problematic of production. Stated simply, the first law of thermodynamics, the law of conservation, posits that matter and energy may be converted and exchanged but can neither be created nor destroyed. As Henry Adams puts it in *The Degradation of the Democratic Dogma*, in which he details and applies a thermodynamic theory of history, there is "incessant transference and conversion," but "nothing was created, nothing was destroyed."[11] Conversion without creation is the thermodynamic conception of force.

The significance of such a conception of force is not hard to detect: the opposition of conversion to creation forms the basis of late nineteenth-century theories of production. In his *Theory of the Leisure Class* (1899), for instance, Thorstein Veblen, not unlike Norris, offers essentially an agrarianist contrast between "industry," which involves "the effort to create a new thing," and the nonproductive or wasteful activities of what Veblen calls "exploit" and "conspicuous consumption," which he defines as "the conversion to his own ends of energies previously directed to some other end by another agent."[12] Such a difference between creation of a new thing and mere conversion of energy, Veblen immediately adds, "coincides with a difference between the sexes."[13] This "coincidence" figures centrally in what is certainly the most familiar American naturalist treatment of the relation between thermodynamics and sexual production, Henry Adams's treatment of the virgin and the dynamo, in *The Education of Henry Adams*.

Adams's text repeats Veblen's opposition of industry and conversion, but with inverse valuation. Adams's meditation on the power of the Virgin takes place, of course, in the Gallery of Machines at the Great Exposition at Paris in 1900. Adams relates that he "ignored almost the whole industrial exhibit," attending instead to the steam engine and dynamo. What these machines perform is a conversion of matter and energy: "To him, the dynamo was but an ingenious channel for conveying somewhere the heat latent in a few tons of poor coal." This power of "interconversion of forms," "endless displacement," and ceaseless "exchange" defines the "wholly new" force of the dynamo. This new power is also perfectly in line with what Adams represents as the final giving way, in the 1890s, of a "simply industrial" economy of production to a "capitalist system" and "machine" ruled by the laws of conversion and exchange.[14]

On the levels of the machine and the body both, Adams's economy is thermodynamic. As he concisely states in his "Letter to American Teachers of History" (1910), "man is a thermodynamic mechanism."[15] And Adams's treatment of the reproductive power of the Virgin offers not finally an alternative to but an extension of this dynamic. The Virgin "was reproduction," Adams observes, "the greatest and most mysterious of energies." She is also "the animated dynamo," the living generator. If the dynamo is an "ingenious channel" of conversion, the Virgin is a "channel of force."[16] The Virgin, not unlike Dreiser's Sister Carrie, who explicitly represents, at the close of the novel, a "medium" and "carrier" of force, and not unlike Norris's "mother," the conveyer of the force that brought men into the world, is a thermodynamic mechanism, a converter of power. In Norris's formulation, "Nature was, then, a gigantic engine" (2:286).

This is enough to indicate, at least provisionally, part of what I suggest is at the back of Norris's rewriting of production and some of the implications and effects of that revision. But as I earlier indicated, the logic of the passage I began by considering is yet more complicated. The translation of earth-mother into force is immediately succeeded by yet another explanation of generation, an explanation that goes even a step further. As if the leveling appeal to an abstract and disembodied "FORCE" fails to reinvest the role of men in production, this account of force is followed by a startling reembodiment. This new and miraculous body recovers not merely a male power of production but also projects the autonomy of that power. The "almighty machine" that displaces the colossal mother is a channel of energy. This machine, however, takes on a strikingly different form: it is a generator of "primordial energy flung out from the hand of the Lord God himself, immortal, calm, infinitely strong." One might say that creation, in Norris's final explanation, is the work of an inexhaustible masturbator, spilling his seed on the ground, the product of a mechanistic and miraculous onanism. The third term in Norris's triptych of mother, force, and onanist-machine places power back into the hands of the immortal and autonomous male technology of generation.

Such an invocation of an autonomous and masturbatory economy of production characterizes the discourse of naturalism generally, and anticipates, for instance, Dreiser's portrayal, in the "trilogy of desire" novels, of the powerful financier, Frank Cowperwood, whose motto—"I satisfy myself"—and notion of procreation—"he liked . . . the idea of self-duplication"—closely resemble the attributes of Norris's god.[17] But it is necessary to note that the very multiplicity of accounts of production that Norris offers, the familiar sublimations and personifications, the "miraculous" explanations of the "miracle of creation"—all indicate the unstable displacements of production in Norris's account. They indicate as well how this very multiplicity and instability may ultimately function as a flexible and polyvalent textual mechanism of relays, conversions, and "crisis" management—as, in fact, a thermodynamic that forms part of the textual mechanism itself.

Moreover, it is necessary to note that the three-part rewriting of reproduction that I have focused on is abruptly interpolated into the "plot" of the novel. The question remains: How does the autonomous and miraculous or "miraculated" production traced here operate in the production and genealogy of Norris's narrative itself?[18] Or from a somewhat different perspective, how does such a management of production structure the naturalist narrative machine? I take up this question in some detail in the following discussion of Norris's *Vandover and the Brute*. But I want to close this account of *The Octopus* by suggesting at least some of the more local "stories" that constitute the novel's techniques of generation and invention of a technology of generation that is also a technology of power.

Put simply, what unites these stories is the desire to project an alternative to biological reproduction, to displace the threat posed by the "women people" (the reduction of men to "mere animalcules" in the process of procreation) and to devise a counter-mode of reproduction (the naturalist machine). One indicator of such a desire is the novel's scandalized representation of "that dreadful substance that was neither solid nor fluid," the scandal posed by "seed," in all its forms, in the novel. The aversion to both women and seed is clearest in the figure who is, interestingly enough, the novel's primary exponent of the technocratic New Agriculture, the wheat-grower Annixter. What above all characterizes Annixter is, on the one side, his horror of what he calls "female girls," and on the other, his horror of a substance he calls "sloop," a "thick, gruel-like, colorless mixture"—"slimy, disgusting stuff"—that he finds in his bed (1:98, 117). Annixter's "hereditary" aversion to semen, and his singular way of cursing—"he is a *pip*"—together with his uneasiness about the "idea" of the woman, make his anti-biological bias hard to miss. The "solution," in the terms that the novel presents, would be a nonbiological and miraculated production that circumvents these threats and projects an autonomous (and male) technique of creation. Such a solution is most readily apparent in the two "love stories" in the novel: Vanamee's romance with Angéle, I and II, and Annixter's romance with his milkmaid, Hilma.

The sheer perversity of the Vanamee-Angéle story provides an almost diagrammatic instance of what might be called the double discourse of the novel, a double writing by which (re)production is displaced or disavowed and rewritten in another register. On one level, the "facts" can be quickly summarized. The shepherd Vanamee meets his sixteen-year-old lover Angéle each night in the churchyard. One night he is anticipated by a mysterious stranger, Angéle is raped, and dies in giving birth to the consequent child ("her death at the moment of her child's birth" [1:45]). Sixteen years later Vanamee resumes his love affair, with Angéle's daughter and exact double. What is most perverse about the story, however, is not its violent and even murderous treatment of sexuality and procreation, but the violence of its symmetries and the urgent translation of these events into the miraculous terms of what Norris calls "romance."

In his representation of "force," Norris translates the "mystery of creation" as the "miracle of re-creation" (2:343), a revision that converts life and death into mere epiphenomenal embodiments of force, cyclical and substitutable repetitions, reproduction as miraculous reproduction. Similarly, Vanamee's mysterious and telepathic calling back of Angéle from the dead is a miracle of re-creation, a simple replacement of daughter for mother: "Angéle or Angéle's daughter, it was all one with him. It was She" (2:106). What such a replacement achieves is a circumvention of both the "defilement" of Angéle's sexuality and the "birth" of her daughter. Vanamee's "mysterious" power of telepathy symmetrically counters what is repeatedly described in the novel as "the mystery of the Other" (1:45), a mystery that ultimately points not merely to the otherness of the unknown rapist but also to the alterity of sexual procreation in the novel. Stated simply, the calling forth of the new Angéle—from, of course, the Seed Ranch—translates reproduction into reincarnation, and such a reincarnation collates exactly with the logistics of production I have already traced. "Angéle," Norris writes, "was realized in the Wheat" (2:347). In his manuscript notes for the novel, Norris is even more precise: "Angéle is the wheat."[19] One might say that Angéle, like the wheat, grows herself, though—again like the wheat—her manifestation or delivery is in other hands. If the "trembling" steam plow "gripping deep into the brown, warm flesh of the land" "seemed to reproduce itself in [Vanamee's] finger-tips" (2:125, 124), Vanamee's reincarnative power offers finally an emphatically personified and personifying technique of nonbiological and autonomous reproduction, or what amounts to a mechanical reproduction of persons.

Hilma Annixter (née Tree) presents a rather different problem for the novel's reworking of generation, since Hilma represents above all a power of maternity, albeit a power miraculated as "the radiance of the unseen crown of motherhood glowing from her forehead" (2:215). The threat that Hilma poses is apparent. Not only does she possess the typically Amazonian body of Norris's mother-women, she also exerts an unsettling control over Annixter. Exercising the "influence of a wife, who was also a mother" (2:209), Hilma literally reforms Annixter, and in a manner that, as should by now be clear, makes for a dangerous vulnerability: "She's made a man of me," Annixter explains, "I was a machine before" (2:180). Not surprisingly, the scene of Annixter's death is juxtaposed to the scene celebrating Hilma's maternity.[20] But if Hilma dutifully miscarries after Annixter's murder, I suggest that the novel's strategy for containing and appropriating the power that Hilma represents takes a somewhat different and more surreptitious form.

Norris's description of Hilma, and more particularly, of her hair, which "seemed almost to have a life of its own, almost Medusa-like" (1:80), reinforces the sense of the milkmaid's unmanning fecundity.[21] But the description invokes as well another and startling cluster of associations: "Deep in between the coils and braids it was of a bitumen brownness, but in the sunlight it vibrated with a sheen like tarnished gold" (1:80). Hilma's hair, bitumen and gold, is something of a mine. What relation might there be between mining and the logistics of generation in the novel?

The Octopus at several points contrasts mining and farming, and in terms directly related to its account of generation. Whereas farming, as we have seen, involves "the elemental passion of Male and Female," mining has a different character. As Annixter protests, "Derrick [the former gold-miner turned rancher] thinks he's still running his mine, and that the same principles will apply to getting grain out of the earth as to getting gold" (1:26). The difference between getting grain and getting gold is a difference between those who "husband" (1:61) their resources and those who "had no love for their land . . . were not attached to the soil" (2:14). And the difference between husbanders and miners is the difference between a loving "embrace" of the earth and violent extraction, "get[ting] the guts out of your land" (1:26), or as Norris depicts mining in *McTeague*, reaching into "the very entrails of the earth . . . boring into the vitals . . . tearing away great yellow gravelly scars in the flanks of them, sucking their blood, extracting gold."[22]

For Norris, as he represents it in *McTeague*, the body of the earth is at once female and "untamed": "she is a vast, unconquered brute . . . savage, sullen, and magnificently indifferent to man." "But," he immediately adds, "there were men in these mountains" (213), and the activities of these men, explicitly allied to the technologies of mining, to the force and "black smoke" of the steam machines, constitute a two-fold assault on the mother-earth by her "progeny": a violent ingestion and assimilation and an extraction from her "vitals" that resembles a violent obstetrics. Nor is Norris's representation of mining at all anomalous in this period. In his popular history of California, *The Sunset Land* (1870), for instance, the Reverend John Todd, a minister who also wrote student manuals governing sexual behavior, contrasts mining, which he calls "the unnatural creation of property" to the more natural creations of agriculture. He notes as well that the eagerness of newly arrived California gold-seekers was so great that "infants were turned out of cradles, that the cradles might be used for washing gold." Todd may be confusing the miner's sieve, called a cradle, with the infant's cradle, but the confusion itself is perhaps significant. (Todd, we may note, officiated at what was called the "wedding" of track that formed the first transcontinental railroad, and thereafter wore a wedding ring made from the gold of the linking "golden spike.") Such a confusion is perhaps shared by Norris's Trina McTeague, for instance, who weeps over the empty bag and box that had contained her gold coins "as other women weep over a dead baby's shoe" (198). Todd, among others, frequently sees the extraction of wealth from the "bosom of mother earth" as a counter-image of procreation, and indeed contraposes such male "unnatural" creation to a threatening female generativity.[23]

The contrast between mining and farming provides a final instance of the novel's rewriting of production. The extraction of gold from the very entrails of the mother-earth is ultimately a species of obstetrics that can dispense with the women-people, and indeed with the body and its dreadful substance, altogether. Seen this way, gold-mining extracts value directly from what Marx called the "the womb of capital itself."[24] And *The Octopus*, making capital of its instabilities and exigencies, provides a virtual map of the crises of production in the late nineteenth century, and of the representations invented to manage these crisis.

Perverse Accouchements

I have to this point deferred considering the aesthetic consequences of the naturalist mechanics of generation I have been indicating. The insistent displacements and ceaseless conversions of force in Norris's texts might be seen to register a thermodynamic technology of power. What remains to be considered are the ways in which such a technology is reinvented and promoted by the techniques of the naturalist narrative itself, and the consequences, at once aesthetic and political, of such an assimilation of the practices of the body in a general and comprehensive physiology and technology of power.

The production of works of art, throughout Norris's texts, takes the form of a process of gestation. In *The Octopus*, for instance, the poet Presley sees his projected work, "germinating from within" (1:6), as a sort of pregnancy: "the desire of creation, of composition, grew big within him. . . . Not for a long time had he 'felt his poem,' as he called this sensation" (1:42). But if such a connection between composition and gestation is familiar enough in romantic and post-romantic aesthetics, what makes for the ultimate *in*effectuality of Presley's art is the novel's invention of a counter-principle of reproduction. Presley's art fails not merely because, as the railroad titan Shelgrim points out, his poem—about labor—simply imitates the painting he "took the idea from" (2:283) rather than directly reproduces "life," but also because Presley's art cannot compete with the technology that Shelgrim represents: "Again and again, he brought up against the railroad, that stubborn iron barrier against which his romance shattered itself to froth and disintegrated, flying spume" (1:10). Presley's labor remains "abortive" (2:85)—filled with "terrible formless shapes, vague figures . . . monstrous" (1:8)—or simply onanistic, because his romantic "spume" cannot compete with the inexhaustible and regulated "white jet" of the steam engine. What the naturalist aesthetic requires, then, is a principle of generation that incorporates rather than opposes the machine: in short, a mechanics that forms part of its very textuality. The discovery and operation of such a machine is the subject of Norris's *Vandover and the Brute*, a novel written before *The Octopus*, but not published until 1914, and a novel centrally about processes of generation, and more particularly, degeneration.

Early on in the novel, the young Vandover looks through the books of his father's library for the dollar bill that the "Old Gentleman" had "at one time misplaced between the leaves of some one of the great tomes."[25] What Vandover finds instead is a long article on "Obstetrics" in an old *Encyclopaedia Britannica*, an article "profusely illustrated with old-fashioned plates and steel engravings. He read it from beginning to end" (10). A little later on, he finds "in the same library" a "Home Book of Art," illustrated with sentimental and idealized pictures of girls and woman, "ideal 'Heads' " (12–13). This juxtaposition, and network, of obstetrics, capital, and a certain style of art initiate and direct the "plot" of the novel.

For one thing, the novel foregrounds an exact equivalence between Vandover's "reckless spending" of the property he inherits from his father and the accelerating process of his degeneration into "the brute." But what at first might appear as a simple "opposition" between proper saving, on the one side, and spendthrift dissipation, on the other, turns out to be somewhat more complicated. Vandover's money, and in particular the bonds or "4 per cents." represent the regulated productions of what is indeed a womb of capital: the bonds, "faithfully brooding over his eighty-nine hundred in the dark of the safety deposit drawer, would bring forth their little quota of twenty-three with absolute certainty" (171). Far from being simply at odds with such a process of production, Vandover's degeneration, precipitated by his discovery of the obstetrics article, involves not merely his reversion into the brute but more precisely another and monstrous process of gestation. If "little by little the brute had grown" (215), this growth, "knitted into him now, fibre for fibre" (277), swelling within the core of his body, "that fatal central place where the brute had its lair" (219), and "feeding its abominable hunger" on "his very self" (30, 316), is a strikingly perverse case in obstetrics. I am suggesting then that if these two economies—of saving or conserving and of dissipating or degeneration—seem to be directly at odds, they are in fact linked by the novel's account of generation. I want to suggest further that these two economies constitute the two governing principles, at once linked and contradictory, of a more general economy of reproduction in the novel, a general economy that functions, moreover, precisely by way of its contradictions. What makes this contradictory economy *operational* is the complex force that Norris calls "the brute."

Vandover sets against the monstrous generation of the brute the "desire of art [that] had grown big within him" (112). The artistic "new life" (116) he desires, as the circumstances of the discovery of the "Home Book of Art" imply, is a projected counter to generation and degeneration both—"Vandover the true man, Vandover the artist . . . not Vandover the lover of women" (112). But Vandover's artistic new life reinvents, on several not entirely compatible levels, the very obstetrics he attempts to evade. Turning to his art as an idealized mode of generation, Vandover discovers in horror that the lines he draws have "no life," that in fact these lines were "those of a child just learning to draw" (224). What "grew under his charcoal" instead were "grotesque and meaningless shapes, mocking caricatures" (225):

> Once more certain shapes and figures were born upon his canvas, but they were no longer the true children of his imagination, they were no longer his own; they were changelings, grotesque abortions. It was as if the brute in him, like some malicious witch, had stolen away the true offspring of his mind, putting in their place these deformed dwarfs, its own hideous spawn. (229)

Vandover explicitly sees this "death" of his art as an abortion, or as the "death of a child of his," and the agent of death or abortion, this midwife or witch-mother, is the very brute gestating within him. The monstrous gestation of the brute is at the same time an obstetrics-in-reverse. Not merely does Vandover's sketch resemble that of a child learning to draw, but his degradation is literally a reversed generation. The growth of the brute is also a return to the scene of birth—"he had become a little child again . . . still near to the great white gates of life" (214). And if Vandover turns to his art as an attempt to "deliver himself by his own exertions" (219), this attempt at self-delivery, in the novel's logic of generation, places Vandover in the places at once of mother, fetus, and obstetrician.

The painting that Vandover is attempting is a sentimental death scene, "The Last Enemy," and his art is still under the influence of "the melodrama of the old English 'Home Book of Art' " (64). But his aesthetic is usurped and twisted by the malicious agency of the brute, and the huge (and "life-size") canvas on which he attempts to body it forth—"the stretcher, blank, and untouched" of "heavy cream-white twill" (223)—becomes the scene of a perverse *accouchement*. Finally, and above all, the brute itself embodies not merely a counter-principle of generation, but a counter-aesthetic as well: an aesthetic of caricature, monstrosity, and deformity, an aesthetic of genesis as *de*generation—that is, the aesthetic of the naturalist novel. Stated as simply as possible, the brute is the generative principle of naturalism.

From one point of view, *Vandover and the Brute* maps a process of degradation; from another, a process of generation. What links these apparently opposed processes is the agency of the brute. A conservative aesthetic of regulation is embodied in the central *objet d'art* in the novel, Vandover's "famous stove," a little machine decorated with pictures depicting the punishment of excess and dissipation, which he carefully tends and adjusts, "the life and soul" of the place (182). Incorporating a regulative aesthetic with a thermodynamic apparatus, the stove might be taken to emblematize, like his father's safe, the ethics of regulation and conservation that Vandover opposes to the brute.[26] But as I have already suggested, such an opposition between conserving and regulating, at one extreme, and the dissipations of the brute, at the other, cannot finally account for the contradictory force that the brute represents. Vandover's resistances to degradation effectively end with the wreck of the steamship *Mazatlan*, a wreck that forecloses his attempts to escape his degeneration. But if, like Vandover's stove, and the "steady" and "cheerful" smokestacks that mark the topography of Norris's (and also Zola's) novels, the steam engine represents a regulated economy of power, the engine is at the same time a "strange huge living creature," an "enormous brute," and its ultimate wreck, the "death" of the brute (135). It appears that the brute is at once a principle of dissipation and of generation—a principle of death and also the life and soul of the novel. The brute operates a contradictory aesthetic allied throughout to the workings of the machine. A closer examination of the "laws" of that machine can perhaps help to clarify the rules of this double discourse and the way in which these contradictions function within a larger system of regulation.

One of the most striking indices of the naturalist aesthetic, as we have seen, is just this close link between generation and degradation, or, more simply, between reproduction and death. If such a linking seems somewhat startling, it in fact functions as one of the basic assumptions of late nineteenth-century theories of reproduction. In *The Octopus* the "return" of Angéle clearly invokes such a connection of generation and dissipation: "Life out of death, eternity rising from out of dissolution . . . the seed dying, rotting, and corrupting in the earth; rising again in life . . . *that which thou sowest is not quickened except it die*" (2:106). Norris's argument here most likely draws on the work of Joseph Le Conte, the religion-oriented evolutionist whose class Norris attended at the University of California at Berkeley in 1892–93. In his essay, "Correlation of Vital with Chemical and Physical Processes" (1875), for instance, Le Conte's thesis is that all life is "generated by decomposition." The seed, for example, "always loses weight in germination; it *cannot* develop unless it is in part consumed; 'it is not quickened except it die.' " In short, Le Conte maintains that "the law of death necessitates the law of reproduction."[27]

The laws Le Conte appeals to here are thermodynamic, and the economy of decomposition without loss he invokes is an explicit mapping of the laws of conservation and of dissipation onto vital processes, a direct application of thermodynamic conceptions of force to the body. As Le Conte repeatedly observes, "the prime object in the body, as in the steam-engine, is [transfer of] force."[28] One consequence of such an application of thermodynamic to vital processes in the late nineteenth century was a radical reconceptualization of the biology of sexual reproduction. This rewriting involved not merely a correlation of genesis and degeneration, but also the invention of a model of sexual difference based on thermodynamic concepts. The deployment of this model is perhaps clearest in the work of the Scottish biologist and sociologist, Patrick Geddes.

In the highly influential *The Evolution of Sex* (1889), Geddes and J. Arthur Thomson propose a biological explanation for differences between male and female social and ethical roles, an explanation based on what they see as an absolute difference between the cellular biology of the sexes. Opposing Darwin's explanation of sexual differentiation, Geddes and Thomson hold that, for instance, "males are stronger, handsomer, or more emotional," not through a process of natural selection, but "simply because they are males." What differentiates male and female is the cellular metabolism that predominates in each. Male cells are "katabolic," that is, characterized by expenditure and breakdown, whereas female cells are "anabolic," conservative and constructive. "Males live at a loss . . . females, on the other hand, live at a profit." By this logic, males embody the second law of thermodynamics—the law of dissipation; females the first—the law of conservation. A thermodynamic biology thus underwrites a political economy of sexual difference, and, more specifically, a typology of genders that precisely corresponds to naturalist typologies of character—for instance, the opposition between McTeague's reckless spending and Trina's "instinct" for saving, in Norris's *McTeague*. But Geddes and Thomson's account is perhaps most interesting for our purposes in the rather remarkable consequences of their theory for the explanation of reproduction they offer. Geddes and Thomson, like Le Conte, focus on the "close connection between reproduction and death," and what follows from this close connection is the notion that "the two facts of reproduction and death . . . may both be described as katabolic crises." And since reproduction in this account is a katabolic process, an unlooked-for dividend that this thermodynamics of the body ultimately entails is that generation—for Geddes as for Norris—is by definition a male process.[29]

These evolutionary accounts of generation help to clarify what I have called the double discourse of the brute in the naturalist novel, the manner in which apparently conflicting processes of generation and degradation operate within a more comprehensive technology of regulation. More generally, these accounts point to the late nineteenth-century double discourse by which the "contradictory" registers of the body and the machine are "floated" in relation to each other and coordinated within what looks like a general economy of power. I have elsewhere attempted to outline how such a "system of flotation" and conversion functions in late nineteenth-century social and novelistic discourses and practices, and the manner in which a circuit of exchange is established between, on the basis of and by way of, conflicting and differentiated practices.[30] What is gradually elaborated is a more or less efficient, more or less effective system of transformations and relays between "opposed" and contradictory registers—between public and private spaces; between social norms and private values; between work and world on the one side, and home and family on the other; between, more generally, "the economic" and "the sexual." A flexible mechanism of adjustment is established, intrinsically promoting a coordination of conflicting practices, while strategically preserving the differences between these practices. Conflicts are, in principle, conscripted into a "circular functionality" between, for instance, "the two registers of the production of goods and the production of producers (and consumers)."[31] Or, in terms of the naturalist logistics I have been considering, between the sieve-like registers of the machine and the body. These new, or rather, newly inflected, strategies of regulation advertise the differences between public and private, and between economic and sexual domains, even as they reinforce and extend the lines of communication between them. But if each appears as the alternative to and sanctuary from the other, as the privileged site from which the other may be criticized and abjured, what these deployments of difference effectively obscure are precisely the links and relays progressively set in place "between" these opposed domains.

From this perspective, the utility and spreading of the thermodynamic model of force in the later nineteenth century becomes more intelligible. This scientifically sanctioned and flexibly generalizable model provided at once a system of transformation and exchange (a principle of conversion) and, in the relays, shifts, and contradictions that facilitate these exchanges, a system of crisis-management (a deployment of difference). The discourse of thermodynamics provided a working model of a new mechanics and biomechanics of power. Moreover, I have been indicating that the transformational system that manages, and capitalizes on, these differences and conflicts between the sexual and the economic, between the body and the machine, is that field of practices that Michel Foucault has called the "biopolitical." Taking as its field of analysis a politics of the body and of the social body, such an analytic identifies a network of practices located "between the empty gesture of the voluntary and the inscrutable efficiency of the involuntary," and reexamines "the endless cleavage between politics and psychology" by focusing on the constitution of the subject as the subject of power.[32] What such an examination of the biopolitical dimension reveals is the subject's disposition at the point of intersection of sexual and political practices and techniques; and what such a production of producers involves is not an ineradicable antinomy between "system" and "subject," between political economy and individual psychology, between anonymous technologies of power and gender-differentiated sexual "identities," but rather a set of exchanges operating between and by way of these antinomies, "choices," and differences. The point, finally, is not to collapse these differences, but to examine their mobility and also their tactical mobilization.

More specifically, I am suggesting that Geddes's theory of sexual difference culminates an attempt, taking off in the late eighteenth century, to provide a biological explanation and justification for gender inequalities. "What was decided among the prehistoric Protozoa," Geddes simply remarks, "cannot be annulled by an Act of Parliament."[33] But most significant in Geddes's account is the perfect fit between the *scientia sexualis* he provides and the power play it entails, the ways in which a thermodynamic technology of power invests and confiscates processes of life and of generation. Such a political economy of reproduction is a local move in a far more thoroughgoing movement over the course of the nineteenth century, a movement that involved the invention and dissemination of a regulative biopolitics. As the Reverend Josiah Strong observed in his popular *The Times and Young Men* (1901), "Evidently, getting the most good out of life, which is getting the most service into it, raises the problem of *The Body*."[34] The aim, as Strong observes elsewhere, is a "balance of power," a balance, as Strong's formula of "getting" and "servicing" indicates, between an ethos of production and an emergent culture of consumption. What is desired is an adjustment between conflicting practices, between a proper conservation of vital powers and the productive utilization of those powers. "The means of self-gratification," Strong insists, "must not outgrow the power of self-control," and more powerful technologies of production must be "accompanied by an increasing power of control."[35]

The focus of this practical and theoretical search for a balance of powers is "the body." One might point to the "medicalization" of late nineteenth-century American society, the rise of therapeutic practices, of eugenics, euthenics, "scientific motherhood," and "physical cultures," and also the rewritings of sexual biology and reproduction already considered. One might cite also William James's investigations of psychic and physical energies and economies, or the evolutionary biology of G. Stanley Hall, whose major work, *Adolescence* (1904), posits a curriculum for the psychic and physical growth of the male adolescent, a process of growth that draws on but ultimately moves beyond the nurturing powers of a miraculated "all mother." (Indeed, Hall's maternal "moon goddess" and her displaced role in generation and growth provides a virtual abstract of the "miraculated" generation that we have examined.) The development of these tactics for dealing with "the problem of the body" has recently begun to be documented, and I want for the moment merely to indicate the larger field in which the technologies of generation I have focused on function.[36] What I want to consider finally, and from a somewhat different perspective, are the ways in which the evolutionary theory allied to these technologies extends such a politics of the body and, collaterally, the ways in which the naturalist novel itself relays these tactics of control.

The Mechanics of Writing

In *Surveiller et punir*, Foucault suggests that the "two great 'discoveries' of the eighteenth century—the progress of societies and the geneses of individuals—were perhaps correlative with the new techniques of power." Such disciplinary techniques, he argues, require bodies at once regulated and productive, and operate, in part, by making use of and assimilating the "genesis" and evolution of the individual within a general tactic of subjection. The discoveries of evolution and individual genesis make possible the articulation of a practice of domination that involves a "new way of administering time" that is also a new way of administering individuals, bodies, and populations. What emerges is an "organic" economy of control linked to and taking hold of the organic evolution of individuals and societies alike. Thus, Foucault suggests, " 'Evolutive historicity,' as it was then constituted—and so profoundly that it is still self-evident for many today—is bound up with a mode of functioning of power," a power that takes as its field of application "the 'dynamics' of continual evolutions."[37]

One of the social practices that underwrites such an administration of power in duration is the nineteenth-century novel, and more particularly the realist novel. The subject of the realist novel, stated very generally, is the internal genesis and evolution of character in society. The realist novel, through techniques of narrative surveillance, organic continuity, and deterministic progress, secures the intelligibility and supervision of individuals in an evolutionary and genetic narration. The linear continuities of the novel make for a "progress" that proceeds as an unfolding and generation of character and action that are always, at least ideally, consistent with their determining antecedents.[38] The naturalist novel involves a mutation in these techniques that consists also in a systematic and totalizing intensification of their effects. This mutation, again stated very generally, makes for functional shifts in emphasis—thematic and narrative shifts, for instance, from inheritance to heredity, from progress (as evolution) to recapitulation (as devolution), from histories of marriage and adultery to case histories of bodies, sexualities, and populations. Yet these differences themselves emphasize a significant continuity: if the realist novel resembles a time machine, the naturalist novel diagrammatically foregrounds, and maps in high relief, the evolutionary dynamics of this machinery.

Far from resisting the realist premises of genesis and generation, the naturalist aesthetic doctrines of determinism and degeneration systematically render explicit and reinforce these premises and the power-effects inscribed in them. Perhaps the most tendentious instance of such a narrative power play occurs, conveniently enough, near the opening of *Vandover and the Brute.* The novel begins by presenting scattered memories or "memory pictures" that from time to time return to Vandover's consciousness "absolutely independent of their importance" (3). The central picture is a scene in a train station, the locomotive "filling the place with a hideous clangor and with the smell of steam and of hot oil" (5). The scene involves the death of Vandover's mother, a death associated throughout the novel with the force of the steam engine. (This memory picture recurs, for instance, at the novel's pivotal point, the wreck of the steamship.) But if this connection between body and machine, as I have argued, governs Norris's mechanics of generation, Vandover himself "could remember nothing connectedly": "What he at first imagined to be the story of his life, on closer inspection turned out to be but a few disconnected incidents" (3). A few paragraphs into the novel, however, there is an abrupt shift in narrative mode. "In order to get at his life," Norris writes, "Vandover would have been obliged to collect these scattered memory pictures as best he could, rearrange them in some more orderly sequence . . . fill in the many gaps" (5). It is just such a move from disconnected pictures to sequential plotting that the novel at this point achieves, a taking over from Vandover that is also an explicit narrative takeover. This rearrangement in sequence engages the novel's relentless logic of devolution, the organic, genetic, and predetermined process of degeneration that Norris's narrative machine traces. Or, as Norris puts it in an essay on the art of the novel titled, neatly enough, "The Mechanics of Fiction," this move from picture to plot is what allows the "entire machinery to labour, full steam, ahead." The art of the novel, for Norris, is a "system of fiction mechanics"; his analogy for the defective novel is "the liner with hastily constructed boilers." And a proper mechanics of fiction involves the "systematizing" of discrete pictures into a segmented and connected series, into an organized and genetic narration: "The great story of the whole novel is told thus as it were in a series of pictures, the author supplying information as to what had intervened."[39] For Norris, the art of the novel is a mechanics of power.

Norris's mechanics of fiction concisely registers and assimilates the technologies of power I have been considering. The techniques of the naturalist machine, its rewritings of production and deployments of a thermodynamic model, reinvent and relay late nineteenth-century social technologies of power and biopower. The achievement of the naturalist novel lies in the devising of a narrative machine that inscribes these technologies as part of its textual practice. In all, the naturalist novel manages late nineteenth-century "crises" of production by the invention of a flexible and totalizing machine of power. Suspending contradictory practices in relation to each other, and intrinsically promoting a coordination and adjustment of these practices, the naturalist machine operates through a double discourse by which the apparently opposed registers of the body and the machine are coordinated within a single technology of regulation.

Part II

Physical Capital: The Romance of the Market in Machine Culture

Persons and Things

Is not slavery to capital less tolerable than slavery to human masters?
George Fitzhugh, *Cannibals All! or Slaves Without Masters* (1857)

One of the most telling passages in Harriet Beecher Stowe's *Uncle Tom's Cabin* (1852) appears just after Tom has been sold to his final owner, Simon Legree, and just before he has arrived at Legree's nightmarish plantation. The subject of this "middle passage" is what Stowe calls "one of the bitterest apportionments" of slavery—the slave's liability to be sold from a "refined family" from which he has acquired "the tastes and feelings which form the atmosphere of such a place" to the "coarsest and most brutal" master, and sold, moreover, "just as a chair or table, which once decorated the superb saloon comes, at last, battered and defaced, to the bar-room of some filthy tavern, or some low haunt of vulgar debauchery." But it is just the uneasy pertinence of the analogy between slave and chair or table that leads Stowe, at this point, to reassert the absolute difference that forms the more fundamental subject of her story: put simply, the difference between a person and a thing. "The great difference is," Stowe goes on to state, "that the table and chair cannot feel and the *man* can."[1]

What is peculiar about this passage, of course, is not what it says but rather the perverse necessity of stating what ought to go without saying, a perversity necessitated by the peculiar institution's denial of the "great difference" between a sentient person and a thing like a chair. Yet it is just the self-evident status of this difference that Stowe's novel ultimately suggests is possibly somewhat too self-evident. For if the great difference between a person and a thing is that one feels and the other doesn't, then Stowe's not at all atypical description, earlier in the novel, of the motherly Quaker Rachel Halliday's chair presents something of a problem, since that description also violates this difference, albeit from the other side:

> It had a turn for quacking and squeaking,—that chair had,—either from having taken cold in early life, or from some asthmatic affection, or perhaps from nervous derangement . . . old Silas Halliday often declared it was as good as any music to him, and the children all avowed that they wouldn't miss of hearing mother's chair for anything in the world. For why? for twenty years or more, nothing but loving words, and gentle moralities, and motherly loving kindness, had come from that chair. (215)

There is certainly nothing very startling about such everyday animism nor, in midcentury discourse generally, about the opposing of such sentimental personifications to slavery's, and also capital's, reifications. But clearly the emphatic transgression of what Stowe sets up as the founding difference between persons and property poses something of a problem in a novel originally subtitled "The Man That Was a Thing" and centered on representing "the feelings of living property." And although the uncertainty about the relations between persons and things is nowhere more insistent than in a novel about the peculiarities of slavery, such an uncertainty characterizes not merely "sentimental" fictions such as *Uncle Tom's Cabin* but also later nineteenth-century American "realist" and "naturalist" writing generally.

"I never saw so much expression in an inanimate thing before," the narrator of Charlotte Perkins Gilman's story "The Yellow Wallpaper" (1892) recalls about her "plain furniture": "there was one chair that always seemed like a strong friend."[2] There is something of a continuity between Stowe's philosophy of furniture and the personifications of her grandniece Gilman, and the continuity clearly indicates something more than a family resemblance. It indicates one of the central concerns of the American novel in this period: a concern about the status of persons, as subjects and as living property, and, collaterally, a concern about the status of material things, such as chairs or tables or, more anxiously, bodies.

In this chapter I want to investigate this configuration of relations—relations of bodies, economies, and forms of representation—

in the later nineteenth century. My primary focus is Henry James's novel *The American*, a novel that charts, in almost diagrammatic fashion, a set of relays between the bodily and the economic and between forms of representation and forms of personation. My account centers on the novel's representation of what James calls "the commercial person": that is, on James's representation of what might be described as commerce embodied in persons. I am interested in the ways in which cultural and economic practices are physically and materially embodied, personified, and reproduced—in short, how bodies and meanings are correlated in market culture. This coordination of bodies and meanings appears, in part, as a problem of putting persons and values "on paper" and I will take up, in what follows, the novel's version of what writers and economists in the later nineteenth century described as the American "paper system."

But what is involved here is not merely the extraordinary American system of making and representing but also what amounts to the systematic making of Americans: the notion of "The American" as an artifact and product, something mass-produced and reproduced. That is, the coordination of bodies and economics that James calls "physical capital" posits the American as something that can be made. More exactly, it counterposes two very different ways of making individuals. These two models of the individual—the model of competitive individualism, on the one side, and of disciplinary individualism, on the other—might be said to epitomize the rival models and tendencies of market culture and machine culture at the turn of the century. In the next chapter I will be focusing on the formation of disciplinary individuals and "statistical persons." Here I am interested in what these tensions between market and machine measures of persons look like and, more precisely, in what the "survivals" of market culture in machine culture look like. That is, I am interested here in what might be called *the romance of the market in machine culture* and in how the romance of the market in machine culture functions (becomes functional) in the internally paradoxical economy of a "consumer society." Hence I want to begin this chapter by examining these different senses of the individuality of the individual and by examining how they are in effect set in motion and also in communication and collaboration in consumer society.

Typical Americans

Near the close of *The American*, just before Christopher Newman returns for his last, brief visit to Paris, there is a small scene that condenses a rich cluster of the novel's interests. Newman, returning to

New York, "sat for three days in the lobby of his hotel, looking out through a huge wall of plate-glass at the unceasing stream of pretty girls in Parisian-looking dresses, undulating past with little parcels nursed against their neat figures."[3] Newman's positioning as spectator concisely invokes one of the most explicit linkings in the novel—an association of a way of seeing and a desire of acquisition or "glory of possession" (118). The scene's association of girl-watching and window-shopping is scarcely surprising in a novel that conceives of sexual relations in terms of market relations: the marriage market or prostitution. Newman's desire to marry is expressed as a desire "to possess, in a word, the best article in the market" (44) and this desire appears preeminently as a seeing at once appropriative and aesthetic. The commercial, sexual, and artifactual or "cultural" are inextricably conjoined by a certain mode of looking in the novel.

Which is to say that Newman is represented as consumer and that "culture" (a term that appears most often in quotation marks in the novel) appears here as the culture of consumption. But if Newman appears here as the typical consumer, it is not quite accurate to say that Newman's move from America to Europe and from business to pleasure represents a move from producer to consumer—the transition that is most often seen as a move, or "fall," from an ethos of production to the culture of consumption.[4] For one thing, as traced in the previous chapter, one of the radical problems of later nineteenth-century American discourse is the problem of securely *locating* the domains of production and producer. Although Newman holds that "manufactures are what [he] care[s] most about," the commercial person, if he makes anything at all, makes money, and his career as consumer seems less at odds with than a continuation of his career as speculator. As Adam Smith notes near the opening of *Wealth of Nations*, commenting on the emergent nexus of speculation, making, and machines:

> All the improvements in machinery, however, have by no means been the inventions of those who had occasion to use the machine. Many improvements have been made by the ingenuity of the makers of the machine, when to make them became the business of a peculiar trade; and some by that of those who are called philosophers or men of speculation, whose trade it is not to do anything, but to observe everything; and who, upon that account, are often capable of combining together the powers of the most distant and dissimilar objects.[5]

If *The American* sets out what might be described as James's theory of the leisure class (and sets it out much in this idiom of liberal market culture), that theory involves a further *migration* of categories of making, doing, and observing. And that migration is one effect of the spreading of mechanisms of production throughout the social body.

By this logic, the division of labor—the division between the users and the makers of machines—reveals, and deploys, a paradoxical discrepancy between doing and making. Correlating using and doing, on the one side, making and observing, on the other, Smith thus makes making an effect of observing and doing an effect of the machine process. Hence the machine process hesitates what exactly it means to produce some thing. One consequence of the disarticulation of making and doing, invention and use, conception and execution is the becoming-visible of the uncertain agency of production (familiarly understood in terms of the alienation of an essential identity of conception and execution). Another is the incorporation of observing or supervising (for example, watching machines or watching the users of machines) into the process of making or producing itself. By the later nineteenth century, as I will be taking up in some detail in the last part of this study, this implicit understanding of the agency of observation becomes explicit in the understanding of processes of supervision (management) and representation (information-processing) *as* the process of production itself. For the moment it may be noted how, in Smith for instance, the improvements in machinery that disarticulate doing and making, and radicalize an uncertainty about the agency of production, at the same time confer a sort of quasi-magical agency or capacity to observation. Hence it appears as if observing or speculating or just looking had the gravitational force of acting at a distance on, and organizing the powers of, all sorts of objects.

The generalized capacity of "combining together" dissimilar powers and objects, drawing into relation and into equivalence "distant" orders of things such as bodies, capital, and artifacts: this *logic of equivalence* is the "classic" logic of the market and of market culture. In what follows I will be concerned with the appeal of the classic logic of the market but also with the limits of such a logic, or fantasy, of equivalence—in short, with the coming into conflict of the market system and the machine process.[6] The managing of the rival tendencies of market culture and machine culture, I will be suggesting, takes the form of a coupling of bodies and machines in the culture of consumption. Such a biomechanics is one of the central topics in these case studies of "the naturalist machine" and "physical capital," and, in the chapters that follow, "statistical persons" and "the body-machine complex."

James's story of "physical capital" inhabits the logic of market culture and instances at once the appeals and the limits of that logic.

Hence my concern is with what that representation of market culture makes visible, and, more specifically, with the relays between making, representing, and reproducing that structure James's account of the making of "The American." Although Newman, for instance, considers his attempt "to pick up aesthetic entertainment"—whether picking up girls or pictures—to be a "strong reaction against questions exclusively commercial" (301) and an "indifference" to the "Exchange" (303), it quickly emerges that aesthetic and erotic matters are not outside of or opposed to the economic in James's story, but bound up with it, through and through and from the start. The internal relations between forms of "cultural capital" and forms of "physical capital" structure James's account of market culture.[7]

What is at stake here can provisionally be clarified by looking more closely at the scene (that is, the New York hotel scene) played out on both sides of "the huge wall of plate-glass." It is not hard to see that the "unceasing stream" of girls makes for a certain crisis of distinctions. The scene involves not merely Newman's "looking" but also an endlessly generated stream of look-alikes: of "pretty girls" in "Parisian-looking dresses." Clearly, this fashionable stream of imitation Parisiennes serves as a metonym of the traffic of, and in, women that the novel represents. But most striking about the passage is what might be called the "undulating" of the description itself, the wavering between registers of its terms. I am referring not merely to the mass-reproduction of copies—of fashionable imitations—that makes possible this flowing line of pretty girls, nor simply to how the consumerism of these parcel-carrying girls doubles the acquisitive watching of the man on the other side of the window. The specular doubling here is a sort of intermirroring, a multiplying of looking and copying, and such a looking and copying figures the production and consumption of commodities in the later nineteenth-century novel. One might say that the large plate glass works as a looking glass—at once window and mirror—even as it uncertainly demarcates and partitions positions of seeing and being seen, of power and possession, of the subjects and objects of desire. The glass "wall" separates the places of the watching subject and the persons and objects that consolidate, but also multiply and disperse, the position of consumer: Newman window-shops looking *out* through the window, positioned as both subject and object of exchange. The plate glass thus works at once as wall and as transparent, and at least apparently permeable, medium or "screen" of exchange.[8] The stream of pretty girls functions at once as the object and the objectification of a general consumerist vision, a consumerism that here pivots on the "neat figure" of the pretty girl.[9]

But the production of copies operates also in a somewhat different and more startling form in this passage, and in a manner that makes even more explicit the exchanges of sexual and economic registers in the novel. Is there a great difference, in the commercial person's way of seeing, between the bodies of these pretty girls and the "neat figures" of a commercial taking-into-account? And is there not something of a resemblance between the "little parcels" they carry and other sorts of bundles these goods seem to displace, living property "nursed against" these figures? The bodily and the economic, one form of reproduction and another, seem almost interchangeable here, as if they were two ways of saying the same thing.

Stated as simply as possible, *The American* poses the problem of the relation between persons and things by way of a consideration of what James calls the "commercial person." The indecisiveness concerning the status of reproduction, mechanical and biological, and of the relations between bodies and figures is crucially significant in a novel that introduces its main character by way of a description of his body as "physical capital" (18) and in which the "private" domain of the embodied subject appears, alternatively, as a "public" spectacle of the market. For Newman public and private "communicate" in terms of exchanges of money and display. (For example, as James notes in passing, "handling money in public was on the contrary positively disagreeable. . .[Newman] had a sort of personal modesty about it, akin to what he would have felt about making a toilet before spectators. But just as it was a gratification to him to be handsomely dressed, just so it was a private satisfaction to him (he enjoyed it very clandestinely) to have interposed, pecuniarily, in a scheme of pleasure" [197].)

Although *The American*, in such instances, protects the opposition between the personal and the commercial, between private and public, between persons and material objects, it is the permeability of these registers and the commerce between them that the novel effectively enacts. "His book overflows with the description of material objects," James observes in 1875, in a review of Taine's *Notes on Paris*, "of face and hair, shoulders and arms, jewels, dresses, and furniture."[10] What the realist inventory of physical capital is indifferent to, it would appear, is a difference, great or even small, between the embodied self and the material object.

The "realist" novel is, then, everywhere underwritten by anxieties about living property and, collaterally, by anxieties about what counts as a person or subject: "Ideas of persons and things, all had dissolved and lost coherence and were seething together in apparently irretrievable chaos."[11] *The American* plays out this panic about the relation between persons and things, and between the body and capital, in terms of a certain logic, or better, *logistics* of representation. By this I mean not simply that the novel focuses on the "topic" of living property, although this topic clearly preoccupies the novel; nor simply that the novel repeatedly represents scenes of representation, copying, and reproduction, although such an insistence is also clear enough. Beyond that, the novel points to the way in which this topic (living property) and this recurrent scene (of representation or reproduction) are inextricably linked in the realist text. *The American* does not simply represent the problematic relation of persons and objects but represents it *as* a problem of representation, imitation, and reproduction. This is the case not because the novel is "really about" representation but because representation in the novel is *already* about the relations of physical capital. What has made the relations of physical capital almost unreadable, and not merely in James's fiction, is the virtually automatic critical tendency to view the obsessive return to scenes of representation in the realist novel as an indication of some internal and critical or self-reflexive distance between a novel's topics and its techniques. Such an account proceeds as if the representation of a subject (including the subject of representation) in the novel automatically placed that subject, as it were, in quotation marks. But it is just the notion of America as an artifact (Christopher Newman, we are told, is named for the man who "invented" America) and the American as artifactual and reproducible that *The American* makes legible.

The novel in fact begins by invoking forms of copying and reproduction in the context of desires at once commercial, sexual, and aesthetic. And in this scene, as in the scene of the glass wall we have just considered, it is the making-equivalent of these desires that is most striking. Newman, having "taken serene possession" of the scene in the Louvre that opens the novel, observes not merely all the pictures on the walls but also "all the copies that were going forward around them." These copies include, it would appear, the copiers themselves: "innumerable young women in irreproachable toilets who devote themselves, in France, to the propagation of masterpieces" (17). The propagation of copies by young women within the museum again peripherally invokes another kind of reproduction. Just outside the museum, Newman's look is insistently drawn to numbers of "white-capped nurses, seated along the benches," as if assembled in efficient lines, and "offering to their infant charges the amplest facilities for nutrition" (29). The novel juxtaposes one form of propagation and another, each in turn attracting the commercial person's interest: turning from copyists to surrogates, Newman looked "at the nurses and babies . . . Newman continued to look at the nurses and babies" (30, 31). This is, as we will see, the first of a series of instances in the novel in which the female body and mechanisms of representation or reproduction indicate each other.

What is particularly marked in this opening scene is a generalized fascination with imitation and reproduction, and this fascination is part of the more general interest in the making of persons and things that the novel proceeds by invoking. Newman, James observes, likes copies: "if the truth must be told, he had often admired the copy much more than the original" (17). On one level, Newman's fascination with the making of copies is not difficult to account for. If manufactures are what Newman cares most about, his interest in making (a term, in its variants, consistently associated with Newman) can be seen as part of the make-up of the "commercial person." As historian David M. Potter, for instance, observes, very much in the idiom of *The American*, "European radical thought is prone to demand that the man of property be stripped of his carriage and his fine clothes. But American radical thought is likely to insist, instead, that the ordinary man is entitled to mass-produced copies, indistinguishable from the originals."[12] On this level, we might say that Newman likes copies because making—making money and things—is what he cares most about.

But if such a notion of the "ordinary" American is ordinary enough, the problem of distinguishing property and identity seems a bit more complicated. For if Newman appears as something like the typical American, it is not so much the idea of the typical American as it is the idea of the American *as* the typical—of Americans as typical, general, and reproducible—that is most powerfully registered here. The American's fascination with making extends to a fascination with the making of persons: the notion of the individual as something that can be made cannot be separated from the interestingly paradoxical economy of "physical capital" and "the self-made man."

Newman is introduced as an embodiment of the "American type" (18), as an instance of what he " 'represented' " (43). He expresses his "national origin," we are told, with an "almost ideal completeness" (18).[13] Newman's *generic* identity is forwarded by repetition in the initial portrait of the American: what is emphasized is Newman's "*typical* vagueness," his "attitude of *general* hospitality to the chances of life . . . so *characteristic* of *many* American faces," his "look of being committed to nothing *in particular*" (18, my emphases). The character of the typical American, of the American as typical, thus stands in a certain tension with the character of his particularity or individuality. More precisely, the difference between the typical and the particular is mapped onto a certain tension between the American's natural body and what he embodies: "But he was not only a fine American," James writes of Newman, "he was in the first place, physically, a fine man" (18). What then is the relation between the individual and natural body, on the one side, and the general and national body, on the other? How do these differences between particularity and typicality and, correlatively, between the natural and national body work in the culture of consumption? And, beyond that, how might the culture of consumption depend on the management and deployment of these differences?

Consumer society, that is, deploys a *rhythm* of the generic and the individual and this rhythm of the general and the particular (a rhythm perhaps clearest in the double logic of "the special" and "the general" that drives the fashion industry) is bound up with the problem of the body and its representations.[14] Newman's "performance of leisure," at the start of the novel, concisely instances these relays between the natural and cultural body.[15] What is emphasized, on the one side, is what might be called the attenuation or even pathologizing of the natural body. This appears in the form of bodily "fatigue" (17, 28), the body "strained," "dazzled" (17), "listless" (19). As Thorstein Veblen observed, commenting on the honorific status accorded "certain diseased conditions of the body" in the regime of "conspicuous consumption," "under the guidance of the canon of pecuniary decency, the men find the resulting artificially induced pathological features attractive."[16]

But what is emphasized on the other is the antidote prescribed to "the numerous class of persons suffering from an overdose of 'culture' " (29): the antidote of "physical culture" and the remaking of the natural body. Hence what is counterposed to the cultivated *depletion* of the natural body (in this case, the depleting "cultivation of the fine arts" [19]) is the culturing and *rebuilding* of the natural body: the art of body-building, the physical cultures of "cold bathing" and "the use of Indian clubs," of health movements, "muscular Christianity" and "homeopathy" (18).[17] The attenuation or consumption of the natural body, in the culture of consumption, is thus the reverse side of the remaking of the natural body, as the artifact of physical culture. This unmaking and making of the body is thus one version of the naturalist machine, its technologies and its biomechanics.

The notion of "physical capital" depends upon this paradoxical relation between the natural and the cultural body. Although Newman is introduced by way of his body as "physical capital," the "conditions of his identity" (19) seem strikingly immune from culture, in either sense. Far from having to keep fit, Newman displays a perfect fit between physical body and identity, between what he owns and what he is:

> He appeared to possess that kind of health and strength which, when found in perfection, are the most impressive—the physical capital which the owner does nothing to "keep up." If he was a muscular Christian, it was quite without knowing it. (18)

Yet if Newman's "general good nature" (38) seems immune from the unmaking and making of the body in consumer culture, this is because his "nature" so consummately embodies that culture. Embodying the market and possessing himself "in perfection" and "quite without knowing it," there is, in Newman, a perfect identity between the "fine man" and the "fine American." And this logic of equivalence between bodies and objects, between persons and things, and between the passions and the interests defines the logic of the market and of market culture. The logic of the market and of market culture projects a fantasy of perfect reciprocity—the equation of interior states and economic conditions, of desires and goods, that makes up, for instance, the subject of "possessive individualism."

For Newman, such a fantasy of self-identity and self-possession is experienced, typically enough, in terms of an identity between desires and things and hence in terms of an opposition between "individual" wants and "social" constraints: "The world, to his sense," James writes of Newman, "was a great bazaar, where one might stroll about and purchase handsome things; but he was no more conscious, individually, of social pressure than he admitted the existence of such a thing as an obligatory purchase" (66). The conflation of wants and goods (the "sense" of "handsome things") and the opposition between "individual" desires and "social" pressures might be taken to epitomize the common sense of laissez-faire individualism. But such a fantasy of self-identity is clearly in a certain tension with the generic individuality that, we have seen, constitutes the typical American and the American as typical. What is involved here, however, is not so much an opposition between individuality and typicality as the tension between two different accounts of the individuality of the individual.

If Newman is "no more conscious" of social pressures than of obligatory purchases, this is because Newman's dream of a free market in persons and things involves an ideal conformity between self and world: "To expand, without bothering about it . . . to the full compass of what he would have called a 'pleasant' experience, was Newman's most definite programme of life" (67). What Newman opposes to this "programme" of self-expansion is a certain form of cultural "obligation": "He had not only a dislike, but a sort of moral mistrust of uncomfortable thoughts," James observes of Newman, "and it was both uncomfortable and slightly contemptible to feel obliged to square oneself with a standard. One's standard was the ideal of one's own good-humoured prosperity, the prosperity which enabled one to give as well as take . . . He had always hated to hurry to catch railroad trains, and yet he had always caught them; and just so an undue solicitude for 'culture' seemed a sort of silly dawdling at the station. . . ." (67). The difference between self-expansion and squaring oneself thus appears in terms of a difference between the desires of the market (the idealized circuit of "give and take") and the discipline of schedules, measures, and standards (the standardized times and places of the railway system, for instance). It is this difference, and tension, between laissez-faire or *possessive individualism and market culture,* on the one side, and what might be called *disciplinary individualism and machine culture,* on the other, that is perpetually reenacted in the paradoxical economy of consumption.

Such a tension is registered in the complex notion of "the standard" itself: in the difference between the notion of the standard as ideal and extraordinary ("one's standard was the ideal of one's own good-humoured prosperity") and the notion of the standard as the normal, the average, and the measurable ("to hurry to catch railroad trains"). But it would be a mistake to argue that the first notion of the standard simply gives way to the second, or that, along the same lines, the opposition between "the individual" and "social pressure" gives way to a smooth assimilation of "the individual" to the norms of "the system," although certainly the understanding of the individual as individualized by his position within the system, as a part of the social mechanism, is something of a commonplace in the discourse of systematic management. The uncertain individuality of the individual, in consumer culture, keeps steadily visible the tension between self-possession and self-discipline, between the particular (one's standard) and the generic (standard ones). The culture of consumption operates, that is, through a coordination of the desires of the market and the disciplines of the machine process: through a coupling of bodies and machines.[18]

James's American, for example, equates "discomfort" with squaring oneself with a standard, but we recall that, for Newman, one's "idea of comfort was to inhabit very large rooms, have a great many of them, and be conscious of their possessing a number of patented mechanical devices" (78). Hence if Newman draws into relation the obligation to catch trains and "an undue solicitude for 'culture,' " this is because Newman's idea of comfort is inseparable from the standardized numbers, sizes, and technologies of mass-production and mass culture. Moreover, though Newman's desire to marry is framed in terms of a free market in women ("I want to possess, in a word, the best article in the market"), the expression of that desire is inseparable from the standardizing idiom of machine culture (a "first-class wife"[76]):

> "She is exactly what I have been looking for . . . a woman that comes up to my standard" (106) . . . "she was perfectly satisfactory . . . You have been holding your head for a week past just as I wanted my wife to hold hers. You say just the things I want her to say. You walk about the room just as I want her to walk. You have just the taste in dress I want her to have. In short, you come up to the mark." (184).

Newman's wants are thus inseparable from the measures, models, and standards of consumer culture ("just as"). And if such models of desire are generally imagined as precisely opposed to obligatory purchases and standards, Newman's romance is expressed in the terms of what had become, by the late nineteenth century, the general technology of romance: that is, the discourse of advertising. What begins to become audible here, as the inner language of the heart, are something like the generic sentiments and publicly shared private experiences of, for example, the mass-produced greeting card, which is also the popular and powerful idiom of desire in, for example, Dreiser's representations of the sensations of consumption.[19]

Culturalism and Consumerism

The "culture of consumption," I have argued, requires the uncertainties about the natural or cultural status of bodies and persons, and deploys the uncertain relation between the natural body and its representations. The continuing fascination, in recent cultural criticism, with the historicity of bodies and the artifactuality (or fashioning) of persons, cannot be separated from the interests and excitations generated by such uncertainties. The fashioning of typical Americans foregrounds the permeability of the boundaries between persons and things and between the individual and the typical in commodity culture: the "conceptual polarity of individualized persons and commoditized things is recent and, culturally speaking, exceptional."[20] There is, of course, a good deal more to be said about the making of individuals in mass culture (and I will be returning to these problems in the following discussions of "statistical persons," "the aesthetics of consumption," and "disciplinary individualism"). Here I want briefly to indicate some of the central difficulties that have emerged in some recent work on the culture of consumption and to indicate also what are perhaps some of the reasons for these difficulties.

One of the governing tendencies in work on the culture of consumption has certainly been the tendency to rewrite the paradoxical relation between the body and its representations (the paradox of *physical capital*) in terms of a narrative sequence: that is, in terms of a "fall" from the natural and the natural body to their representations or substitutes, put simply, a fall from "the natural" to "the cultural." Hence both cultural conservatives, such as Christopher Lasch, and cultural radicals, such as Stuart Ewen, repeat the story of a fall from "substance" to "style" and from things to images that reiterates the terms of a (Platonic "myth of the cave") condemnation of representation as such, or, at the least, a disturbance in the normative relations between things and their representations.[21] Hence Daniel Bell, in his generally impressive study of *The Cultural Contradictions of Capitalism*, sets out the intrinsically contradictory tendencies of capitalist culture—the tensions, but also the relays, between an ascetics of production (self-discipline) and an aesthetics of consumption (self-aggrandizement)—but nevertheless tends to convert these tensions into a familiar narrative of "transformations from simplicity to luxury," from "asceticism to hedonism," from the needs of production to the desires of culture and consumption. What this amounts to in effect is a rewriting of the cultural contradictions *of* capitalism as a contradiction *between* consumer culture and capitalism, and thus a refusal of both culture and consumption.[22]

Stated simply, the move to "periodize" the fall into consumer culture takes the form of a standard, and remarkably portable, story of a fall from production to consumption (or from industry to luxury, or from use to exchange, etc.) that at times seems as crude as the claim that people grew things in the first half of the nineteenth century and ate them in the second half.[23] The recalcitrance of such crude periodizations intimates that another sort of anxiety or appeal is perhaps being staged here and played out by way of the uncertainties about the natural or artifactual status of bodies and persons. It intimates that the problem of periodizing something like "the culture of consumption" is perhaps inseparable from the general problem of periodizing or historicizing bodies and persons.

For one thing, although the move from an ethos of production to a culture of consumption is often seen in terms of a replacement of the natural body by its substitutes or representations, it's not hard to see that the *return* of the body is a central tendency in consumer culture. This is not merely because, as Werner Sombart noted, in his study of luxury and early capitalism, there is a "high degree of linkage of [luxury goods] consumption to body, person, and personality."[24] It is also because the simultaneous solicitation and disavowal of the natural body in consumer culture is, as we have begun to see, bound up with the "rival" logics of the market and the machine process.

These contradictory accounts of the body and culture are nowhere clearer than in the indictments of "conspicuous consumption," indictments restated, with a striking predictability, from Veblen's *Theory of the Leisure Class* (1899) on. Veblen, for example, understands the return of the body as one of the "archaic survivals" of a nature directly at odds with the disciplines of the machine process and machine culture. The status accorded what he sardonically calls the "radiant body" (that is, the natural body as charismatic, radiating a sort of animistic force) is, for Veblen, one of "the survivals of archaic traits of human *nature* under the modern *culture*"[25] (my emphasis). And for Veblen, this archaic and vestigial attachment to the natural body, its desires and its animisms, is the scandal of market culture and conspicuous consumption, a scandal that persists not merely in the rituals of consumption but in the interpretation of the economy in terms of an idealization of "the market" and the interpretation of the market in "sensuous terms." For Veblen, such an interpretation of economic institutions in terms of "the sensations of consumption" (a version of the hedonistic calculus) or, more generally, in terms of "human nature" (the systemic adequation of the passions and the interests) is directly at odds with the "metaphysics of the machine technology" and the understanding of economic life in terms of the "discipline of the machine process."[26] Veblen's opposition of the market and the machine thus appears in terms of an opposition of the bodily and "the natural" to the technological and "the cultural" and an endorsement of the replacement of the first by the second: the replacement of nature by culture and of bodies by machines.

The animus against the natural body and attraction to the machine process seem clear enough. But the contradictions that Veblen's avowed anti-naturalism issue in are no less apparent, and not least in the utterly contradictory status of "culture" in Veblen's account. Veblen's anti-naturalism would seem to entail a strictly sociological or culturalist account of persons and things (the positing of a sociological and "historical index of all truth"). Yet such an emphasis is directly at odds with the reverse tendency in Veblen's work: its emphatic anti-culturalism and what amounts to, in Theodor Adorno's terms, "Veblen's attack on culture."[27]

Two divergent impulses govern Veblen's account. On the one side, Veblen insists on a thoroughgoing sociologism or sheer culturalism (the social index of meanings and values; persons and things as reflexes of the culture that "produces" them). On the other, he appeals to residual, even instinctual human nature—centrally, "the instinct of workmanship"—in order to ground his opposition to the culture of consumption (what Adorno calls Veblen's "crude belief in the natural"). Hence if Veblen's attack on culture is explicitly the expression of contempt for "the aesthetic" (the aesthetic as mere fashion, as inutility or, simply, waste), it indicates also an anxiety about the very sociology of knowledge and value—the sheer culturalism—that Veblen's account takes as one of its working premises.

This version of the culturalism/essentialism debate, its impasses, paradoxes, and irresolutions, is by now familiar enough. If the governing impulse of recent "cultural studies," for example, has been to insist on the artifactuality of the social and on the sociality of artifacts, it also everywhere remains hesitant about this culturalist tautology. It might be suggested that the recent forms of the culturalist debate register, albeit in the form of a theoretical rewriting, the contradictory status of "the natural" and of "the cultural" in consumer culture. But the recalcitrance of these contradictions is perhaps less striking than the appeals such hesitations generate and the relays they trace across the body, the market, and cultural forms.

The work of the sociologist Pierre Bourdieu, for instance, continues the emphasis on investigating what Veblen calls "intangible assets" and what Bourdieu calls symbolic or "cultural capital." But if such a cultural poetics acknowledges the social force of the intangible, the symbolic, and the cultural—the power of artifacts and representations, for example—it does so finally only by "accounting for" culture as a reflex of interests. Both Veblen and Bourdieu operate as cultural efficiency experts, assessing the interestedness of disinterestedness, the invidious social distinctions enforced by the ostensible disinterestedness of the aesthetic. The imperative is what Bourdieu describes as "a systematic reduction of the things of art to the things of life." Or, as Bourdieu more exactly puts it: "culture depends on the laws of the market."[28]

Such a reduction of culture to the market is, in turn, grounded by another, and somewhat paradoxical, reduction. "One cannot fully understand cultural practices," Bourdieu indicates at the outset of *Distinction: A Social Critique of the Judgement of Taste*, "unless 'culture,' in the restricted, normative sense of ordinary usage, is brought back into 'culture' in the anthropological sense, and the elaborated taste for the most refined objects is reconnected with the elementary taste for the flavours of food."[29] The understanding of culture in terms of the market is, at the same time, an understanding of culture and the market in terms of the natural body. Grounding culture in the market and the market in human nature, such an account seems to suggest that accounting for taste, in the social sense, and accounting for taste, in the bodily sense, are two ways of saying the same thing.

But it is such a logic of equivalence by which the bodily, the aesthetic, and the economic indicate each other, in circular fashion, that conserves the logic of market culture. This is also the set of relays between persons and things, bodies and artifacts, the individual and cultural body that makes up what James calls "physical capital." James's representation of the typical American, and the practice of the later nineteenth-century novel generally, richly instance the uncertain relations between persons and things, artifacts and bodies, and instance also the anxieties and appeals they generate. The problem of physical capital, in brief, cannot be separated from the problem of cultural capital and, more specifically, from the *hesitated* relations between persons and representations, and between the natural and artifactual status of the individual.

The system of flotation by which the individual and the typical, the natural and the cultural, "the real" and its imitations or simulations are at once opposed and coupled, floated in relation to each other and "adjusted" in consumer culture, may by now be clear enough.[30] But if such a system seems to posit a relentlessly *abstract* notion of the typical individual—the abstraction of "the body" and of "personhood"—it perhaps also makes visible the double process of abstraction and embodiment that fashions the generic individuals of commodity culture. I am referring to the double process by which the privilege of relative disembodiment that defines the citizen of liberal market society ("abstract universal personhood") is correlated with the making-conspicuous of the body in consumer society. Opposed on part of their surface but communicating on another level, the privilege of abstraction and the requirement of embodiment are linked together in the consumer *body-in-the-abstract*: in the fashioning of the generic *model-body* of the consumer, abstracted and individualized at once. This is the double process of abstraction and embodiment that articulates the relays between the natural and national body we have been tracing and allows the national and natural bodies to stand in for each other. It is this translation or communication between economic and physical registers that fashions "The American" as something like a brand-name for the model citizen of commodity culture.[31] Newman's social "sense of equality," we are told, "was not an aggressive taste or an aesthetic theory, but something as natural and organic as a physical appetite" (152). "The American" sense of equality thus refers back to the natural body ("physical appetite") which refers back to the market ("physical capital") which refers back to the sense of equality (the abstract equivalences of market culture).

These are the bodies-in-the-abstract that populate consumer and machine culture. But it is not merely that the privilege of relative disembodiment requires the more deeply embodied bodies (in consumer society: the female body, the racialized body, the working body) against which this privilege can be measured. Beyond that, it requires those more visibly embodied figures that, on the one side, epitomize the tensions between the typical and the individual and between the artifactual and the natural and, on the other, are the figures through which these tensions can be at once recognized and displaced or disavowed. And, beyond that, it is just these tensions, epitomized by and displaced upon the overly explicit embodiments of physical capital, that fascinate and arouse interest in consumer culture: for instance, the erotic visibilities of the female body or the "raciness" of the racialized body.

What fascinates and excites James's American are just these tensions and instabilities. What Newman wants as a "first-class wife" seems to epitomize the artifactuality of persons, the making of persons through the fashioning "processes of culture": hence "Madame de Cintré gave Newman the sense of an elaborate education, of her having passed through mysterious ceremonies and processes of *culture* in her youth, of her having been *fashioned* and made flexible to certain exalted *social* needs" (110, my emphasis). But what Newman finds stimulating is not quite Claire's consummate artifactuality, nor what is also seen as her "good nature" but rather the impossibility of securely locating the line that divides her nature and her culture:

> But looking at the matter with an eye to private felicity, Newman wondered where, in so exquisite a compound, nature and art showed their dividing line. Where did the special intention separate from the habit of good manners? Where did urbanity end and sincerity begin? Newman asked himself these questions even while he stood ready to accept the admired object in all its complexity; he felt that he could do so in profound security, and examine its mechanism afterward, at leisure (110)

The "mechanism" behind these questions—that is, the question that takes the form "Does she or doesn't she?"—thus appears (to the male inquirer) as the mechanism of the feminine, or even of the feminine as mechanism. That is, the uncertainties about the body and its representations, about the natural and cultural body, about "culturalism" generally, are here read through and displaced upon the "compound" female body. (Such a radical entanglement of the culturalism question and the question of the feminine also indicates, at the least, one way the interests, and erotics, that inhabit the deeply invested but often highly abstracted debates about culturalism "as such" may be particularized and one way the general question of the naturalness or artifactuality and even machinelikeness of persons is reembodied—questions, in this context, neither reducible to nor separable from such entanglements of mechanism and gender.)

At one extreme, such a notion of the feminine appears in an explicit identification of the woman and the machine. This identification is nowhere more explicit than in the novel's lurid account of the self-made woman, Mademoiselle Nioche:

> "But she is a very curious and ingenious piece of machinery. I like to see it in operation." "Well, I have seen some very curious machines, too," said Newman; "and once, in a needle factory, I saw a gentleman from the city, who had stepped too near one of them, picked up as neatly as if he had been prodded by a fork, swallowed down straight, and ground into small pieces." (179–80)

The rewriting of the familiarly feminine and domestic (a needle, a fork) as a monstrous and mechanical power of consumption is what here registers the prostitute's "ingenious" unnaturalness. There is a rapid and extraordinary conflation, in this passage, of the normatively separate spheres of the home and the factory, of the private, female, domestic space of consumption and public, male space of production. There is an extraordinarily condensed and violent registration, that is, of the later nineteenth-century conflation of domesticity and consumption, and the reorientation of domestic spaces as market places of consumption.[32]

In this extreme version, the mechanism of the feminine takes the form of a Nana-like eating machine or the prostitute-automaton of Villiers de l'Isle-Adam's *L'Eve future* (1886). But it is less this terminal version of the woman-machine than a more everyday version of the question of the feminine and its mechanisms that such moments evoke. It might be argued that what the prostitute violates is precisely the normative link between "the female" and "the natural": the set of analogies positing that "female is to male as nature is to culture." The scandal of the prostitute or "painted woman" would seem to be her *un*naturalness or artifactuality (the artifice that turns biology to economy). But if the painted woman would thus seem to violate the recalcitrant association of the female and the natural, she would thus seem to exemplify the equally recalcitrant, if reverse, association—the normative link between the female and the cultural, the association of "aesthetics and the feminine."[33] Hence neither the identification of the feminine with the natural nor the identification of the feminine with the cultural but, instead, their uncertain mixture—*the miscegenation of the natural and the cultural*—is what incites, at once, panic and interest. Another way of saying this is that what is scandalous about the figure of the prostitute, in the realist novel, is that she embodies, with a violent explicitness, the mixed logic of physical capital: utterly artifactual and utterly physical at once, capital with a human face.

This is perhaps enough provisionally to indicate the entanglements between the bodily and the economic that make up the hybrid form of physical capital. The scandalized representation of the prostitute instances also the *disavowal* of just such an entanglement and, in the relaying of economics as erotics and as the mechanism of the feminine, the paths that disavowal takes. These are the relations between erotics and possession that James (in his preface to *The American*) calls "romantic property" and the relations between property and love that James calls "the romance," and it is to the turn-of-the-century reinvention of the romance and of romantic property to which I now want to turn.

Romantic Property

One way of situating the turn-of-the-century reinvention of the American romance is by looking more closely at how James represents the relation between a "person" and a "picture," and, more generally, between persons and representations in the novel. Newman's admiration for copies points to what James describes as the American's aesthetic "baffle[ment]," a confusion of copy and original allied to "the damning fault (as we have lately discovered it to be) of confounding the merit of the artist with that of his work" (19). It does not really matter whether Newman confounds the work of art with the painting woman or confounds it with the painted woman (in fact, he does both). Either way, Newman instinctively conflates the things of art and the things of life and such a conflation foregrounds the conflation of body and capital that both the self-made man and the made-up woman epitomize. Noémie's painting of the Madonna clearly resembles the making-up of a prostitute ("she deposited a rosy blotch in the middle of the Madonna's cheek" [210]), and the "copy" and the "beautiful subject" she offers to sale implicitly refer, of course, both to her painting and to *herself* as a remarkable and marketable imitation of the type of the "perfect Parisienne."

The confounding of a person and a picture enacted here poses again the uncertain relation between persons and things, this time overtly in terms of the artifact itself. From one perspective (that of a "lately-discovered" formalism), the confounding of a person and a picture is simply a mistake, a "damning fault"; from another, it is just this mistake that, in effect, turns a portrait into a character, into a person defined not by a freedom from representation but, instead, *characterized*, in both senses, by what he or she imitates or represents. The problem is, as Tristram remarks, that in Paris "they imitate, you know, so deucedly well . . . you can't tell the things [original and imitation, person and picture] apart" (28). The novel's internal debate between formalism and realism is thus part of its pressuring of the relations between persons and artifacts, or what amount to the relays between representations and living property.

The most explicit invocation of the link between representation and property in the novel occurs in the second scene in the museum (chap. 11), Newman's meeting with Noémie to check on "the progress of his copies" which is also, in the meeting of Valentin de Bellegarde and Noémie, the "progress of the young lady herself" (129). What centers this scene is not Noémie's copying but, rather, and in response to Valentin's critique of her painting, her defacement of the picture:

> "I know the truth—I know the truth," Mademoiselle Noémie repeated. And dipping a brush into a clot of red paint, she drew a great horizontal daub across her unfinished picture.
>
> "What is that?" asked Newman.
>
> Without answering, she drew another long crimson daub, in a vertical direction, down the middle of her canvas, and so, in a moment, completed the rough indication of a cross. "It is the sign of the truth," she said at last. (133)

The passage proceeds by juxtaposing two different but interdependent accounts of representation and by staging a certain hesitation between them. The defacement of the picture is translated into a mark or sign; the mark that crosses out the portrait becomes a cross, a form of iconic writing, "the sign of the truth."

There is more to be said about this highly condensed account of representation and marking: of inscription as picture or as writing, as mere material "daub" or as sign. The tensions between picture and writing, and between mere physical marks and legible signs, are crucial to the problematic of representation and personation in the novel. Here the painted woman's defacement of her portrait evokes two very different responses, toward persons and toward artifacts. Not surprisingly, Newman resists the marring of what the picture looks like (that is, he resists the dissolving or reduction of likeness and identity to their mechanisms or material), complaining that these crimson marks "have spoiled your picture." But the "cultured" Valentin's very different commentary embraces at once the mechanics of representation and the mechanism of the feminine, in effect equating them: " 'I like it better that way than as it was before,' " Valentin declares: " 'Now it is more interesting. It tells a story. Is it for sale, mademoiselle?' " For Valentin, Noémie's defacement of her portrait is a move from picture to narrative and the move from picture to narrative also and necessarily entails a story of exchange—"Is it for sale?"—and of capital embodied in persons—" 'Everything I have is for sale,' said Mademoiselle Noémie."

What this passage enacts is an almost schematic configuration of relations between commercial exchanges, persons, and stories. It indicates that to eliminate the commercial is to eliminate what counts as a person and to eliminate what counts as a person is to eliminate what, for James, is the very principle of a "story": the exchanges and relations that, as James expresses it in his preface to *Roderick Hudson*, "stop nowhere." The difference between a person and a picture in *The American* is a difference between a story and a portrait, or, in accordance with the names that James gives to these different practices of representation, the difference between the novel and the romance.

What characterizes the romance, as James argues in his preface to *The American*, is a fantasy of unrelatedness: the radical autonomy of persons. The romance, as opposed to the novel, consists of a "liberated" "kind of experience," of "experience disengaged, disembroiled, disencumbered," of experience, above all, relieved of "the inconvenience of a *related*, a measurable state . . . at large and unrelated" (11). If story consists in relations, it is thus precisely a "sacrifice" of story that *The American* risks or, in fact, advertises: the novel turns, in its second half and with the breaking of the engagement, from the entanglement of relations that make up the commercial person to a form of representation that is something like (and indeed literalizes) the disengaged and unrelated state that James compares to "a page torn out of a romance" (276). Hence the turning away from forms of relation and engagement—the turn from novel to romance—is represented as the turn from a story of exchange to a romance exempt from both story and exchange.

Yet if the reversion to romance seems to protect a fantasy of autonomy, such a fantasy involves not a radical opposition between persons and possessions but something like the opposite: the perfect identification of persons and possessions, the fantasy of reciprocity that defines market culture. The nostalgia for the romance, in *The American*, represents a nostalgia for the forms of self-identity and self-possession that James calls "romantic property." I want now to set out what the fantasy of romantic property looks like, first in the midcentury novelistic practice that lies back of James's nostalgia for the romance, and second, in the versions of possession and representation that make up James's turn-of-the-century reinvention of the romance. What I want finally to consider is how what Veblen might call the "archaic survivals" of the romance function in machine culture, and how the romance of the market, and the romance of possession and self-possession, continue to *animate* (again, in Veblen's sense) relations of property and identity in turn-of-the century consumer society and beyond.

On the threshold of the meeting with Noémie that we've just glanced at, Newman remarks to Valentin that "I have in fact come to see a person and not a picture." In his study *Hawthorne*, written shortly after *The American*, James critically comments on the "light and vague" and "disembodied" typicality of the characters of Hawthorne's *The House of the Seven Gables*: "they are all pictures rather than persons."[34] The difference between a person and a picture invoked here is for James also a difference between romance and novelistic forms and between the genre-fictions of the body and personhood these rival forms of representation entail. A brief consideration of Hawthorne's *The House of the Seven Gables* (1851) can perhaps clarify these genre-fictions of bodies and persons and their implications.

It does not take much pressure to see that Hawthorne's novel traces a transition from aristocratic to bourgeois models of economic and familial relations: as Hawthorne concisely expresses it, "the state of the family had changed,"[35] Nor is it hard to see that this transition is bound up with "the problem of The Body." The novel focuses on the body as the locus of both history and meaning: on the body *as* embodiment. Hawthorne's portrayal of the aristocratic Pyncheons' "ancestral" chickens, for instance, economically exemplifies this correlation of history and biology. The chickens function as a "symbol of the life of the old house," "embodying its interpretation," even as they "embodied the traditionary peculiarities of their whole line of progenitors." The smallest of the chickens has "aggregated into itself the ages": all its ancestors "were squeezed into its little body" (151). But if Hawthorne is here sketching a sort of protogenetic fantasy or quasi-biological account of how something like original sin could be passed on from generation to generation, "squeezed into" successive shapes and "reproducing itself in successive generations" (240), the novel's original transgression has to do with property and its transmission.[36] What focuses this nexus of property and the body is the practice of the family.

The small body of the ancestral chicken represents what Hawthorne calls aristocratic "transmission." Phoebe Pyncheon's "nice little body" (79) centers Hawthorne's treatment of the new "republican" family of conjugality and domesticity. And if "transmission"—transmission of property and identity—defines the aristocratic body, what Hawthorne calls "communication," "circulation," and "intercourse" define its bourgeois successor.

It is not simply that Phoebe, through a sort of "homely witchcraft" (72), converts the house of the seven gables into a home, but that this conversion involves a new relation between persons and things and between things like bodies. Phoebe's little body exerts a "force" that has the capacity "to bring out the hidden capabilities of things around [her]" (71). Her "influence" is manifested, for instance, in "the effect which she produced on a character of so much more mass than her own" (137), an effect, at once spiritual and physical, of magnetic attraction. Phoebe's assumption of the office of housewife involves "not so much assuming the office as attracting it to herself, by the magnetism of innate fitness" (76). Above all, what characterizes Phoebe's rearrangement of the relations of persons and things is a power of personification, a "natural magic" that operates as a species of animism.[37] Finally, this circulation or communication established between persons and things takes the form of an eroticization of everyday relations: Phoebe's "human intercourse" (141) animates bodies and objects and alters the very "air" of the house of the seven gables—"she impregnated it" (143).

From one point of view, Phoebe's impregnating intercourse resolves the problematic relation between persons and things by creating what is typically called the "separate sphere" of domesticity and conjugality: the sexualized, privatized space of the new state of the family, the romance of the "new Eden" that "separated Phoebe and [Holgrave] from the world" (305).[38] But from another, the animistic romance of the new family appears not as separate from but in fact as entailing the very "worldliness" it seems to oppose. Phoebe's homely magic is implicitly associated with an apparently very different kind of witchcraft in the novel. The "little circlet of the school-boy's copper coin"—the first income in Hepzibah Pyncheon's cent shop—communicates a certain "galvanic impulse." The little copper, "dim and lusterless though it was, with the small services which it had been doing, here and there about the world, had proved a talisman . . . It was as potent, and perhaps endowed with the same kind of efficacy, as a galvanic ring!" (52). An animating magnetism, a power of circulation and intercourse, characterizes both Phoebe's domestic influence and the witchcraft of the marketplace. From this rather different point of view, what Phoebe's nice little body is an embodiment of is the circulating medium of capital itself.[39]

The sexualization of economic relations appears here in part as a displacement of economic difference onto sexual difference, a displacement that, in Hawthorne as in James, is linked to the relations of narrative romance as such. Hence the transition from hereditary aristocracy to the conjugal space of the bourgeois family is broached in the inset narrative that at once marks and accomplishes that shift: the narration of the story of Alice Pyncheon which, it will be recalled, draws Phoebe and Holgrave together. In short, the "class" confrontation between Alice and Maule, occasioned by a dispute over property, becomes a confrontation between the sexes: "setting aside all advantages of rank . . . Alice put woman's might against man's might" (203). It's this rewriting of economics as erotics, or, more precisely, the relays progressively elaborated between these registers, that defines the circular relation between desires and the market that makes up "romantic property."[40]

The understanding of human nature in terms of the market and the understanding of the market in sensuous terms, in terms of "the sensations of consumption": this tautological relation between interior states and economic conditions (between the passions and the interests) defines the circular logic of market culture. But whatever the status of such a logic in mid-nineteenth-century America—and, according to Alfred D. Chandler, Jr., the "American economy of the 1840s provides a believable illustration of the working of the untrammeled market economy"[41]—it is the recalcitrance of that logic, in what can by no means be described as an "untrammeled market economy" at the turn of the century and beyond, that the reinvention of the romance and of the relations of romantic property makes visible. Two general questions present themselves at this point: First, how do the regression to the logic of the market and the "archaic survivals" of romance and romantic property continue to operate in machine culture and continue to structure its rituals of consumption? And, second, how does the appeal of the market, and the reflexive relation between interior states and social demands that it projects, continue to structure, for instance, accounts of American culture "as such"?

Status, Contract, Discipline

"The recurrent Jamesian subject," Leo Bersani has observed, "is freedom," and it is in the name of such a radical freedom and autonomy of the subject that James, it will be recalled, defends his turn to romance.[42] What most of all characterizes the conditions of Newman's identity is his "freedom" and his autonomous and unrelated state: his "almost ideal completeness" (18). Newman's freedom is centrally a freedom *from* relations, familial and erotic. "And are you perfectly free?," Noémie asks Newman:

> "How do you mean, free?" "You have nothing to bother you—no family, no wife, no *fiancée*?" "Yes, I am tolerably free." (62)

Newman's notion of relation itself is not at all at odds with such a "perfect" freedom. What Newman proposes to Claire is marriage not as a "losing [of] your freedom" (113) but as its perfection: "you ought to be perfectly free and marriage will make you so" (115). This is in part because Newman sees marriage as an extraction from familial relations, expressing not a desire to "come into" the family but rather the desire to "take Madame de Cintré out of it" (144). "I want to marry your sister, that's all . . . I am not marrying you, you know, sir" (143). Newman's republican or "federated" version of alliance posits autonomy and disconnection. For Newman, that is, the parties of an alliance are "very different parties": "I see no connection between you" (250–51).[43]

This twin form of alliance and disconnection—of "different parties"—reinvents, in terms of the romance, what are also the terms, and conditions, of romantic property: the terms of *the contract*. The conflict between aristocratic and commercial persons in the novel instances what is familiarly understood as the conflict between, or transition from, the code of status and the code of contract. In this instance, the conflict between status and contract appears as a conflict over the meaning of promises—in short, the difference between breach of honor and breach of contract. The making or breaking of promises underwrites the plot of the novel, from Newman's "solemn promise" to defer speaking of marriage for six months (115) to his attempt to restore relations by signing a "paper promising never to come back to Europe" (219). For Valentin—"the soul of honor"—the promise is aligned with the aristocratic code of honor (the duel). For Newman, the promise is aligned with the two-sided form of relating different parties, while maintaining their individuality and autonomy (the contract). As the commercial person puts it: " 'If you stick to your own side of the contract we shall not quarrel; that is all I ask of you,' said Newman" (147).

There is a close connection, as the historian Thomas L. Haskell has recently argued, between the emergence of the "norm of promise keeping" and the normative conditions of market capitalism.[44] (Haskell's "topic" is the problem of slavery in the context of liberal capitalist society—that is, the conflict between contractualism and "older" versions of the ownership of persons.) A contract represents, of course, a legal exchange of promises, and, as Haskell traces, the "growing reliance on mutual promises, or contractual relations, in lieu of relations based on status, custom, or traditional authority comes very close to the heart of what we mean by 'the rise of capitalism.' "[45] One consequence of this transformation in the character of relations is a new characterization of the individuality of the individual. The task becomes, in Nietzsche's terms, "to breed an animal *with the right to make promises*"—that is, an individual defined at once by the radical freedom of will that contractual promises presuppose ("I will") and by a power over circumstance and over temporal and physical distance, a standing security for the future delivery on one's will, that the making and keeping of promises requires ("I must").[46] Contract requires, in other words, the "sovereign individual" defined at once by the sovereignty of his will and by an obligation to abjure willfulness. Hence the "possessive individual" of the contract system is (to paraphrase William James) possessed by his possessions. That is to say, what defines the freedom and autonomy of the individual, the *principle of identity and self-identity* over time and circumstance that makes promise keeping possible is the ability, or what Newman calls the "duty," of "resembling oneself" (158). The determination of the individual by his determinations is what makes the duty of self-identity possible.[47]

To "have a character," in, for example, Benjamin Franklin's sense of having and sense of character, is to be a subject who can make good on his promises. That Franklin is invoked, more than once in *The American*, as the archetype of the typical American, suggests that the self-made character of the commercial person is an embodiment of the promissory model of contract, and the forms of possession and of self-possession, it entails. But if Franklin is invoked as Newman's archetype, this is perhaps to suggest also that there is something anachronistic about the understanding of the typical American at the turn of the century in terms of the model of contract and in terms of the contractual logic of self-identity.

From one point of view, the conflict in *The American* is the conflict between status and contract. From another, neither status nor contract, nor their conflict, provides the model for understanding the economic and affective "conditions of identity" or individuality at the turn of the century. Neither the classic notion of the market nor the classic model of contract accurately registers the conditions of identity in machine culture.[48] But for that reason the insistently anachronistic rewriting of those conditions has perhaps its own logic in consumer society.

The invidious distinctions of market culture more closely resemble the honorific distinctions of status than the rank-and-file classifications and disciplines of machine culture resemble either. For Veblen, for instance, it is precisely the reinvention of the invidious and competitive distinctions of "barbarian" and "feudal" culture in the rituals of conspicuous consumption, and the "archaic survivals" of the charismatic and "radiant body" and its fetishization, that above all constitute the regressive "canons of reputability" of the leisure class. These archaisms are, by Veblen's account, simply regressions that have remained behind the "metaphysic" of the machine process, "as a disciplinary factor in modern culture," and that amount to symptoms of a nostalgia for market culture.[49] What Veblen misses in this dismissal of regression as "mere illusion" is, Adorno argues, "the compulsive element in modern archaism." As Adorno expresses it: to Veblen, the railway station built to look like "the phony castle is simply anachronistic. He does not understand the distinctly modern character of regression."[50] And the distinctly modern character of regression, for Adorno, is inseparable from the disciplines of the body-machine complex of modern culture. If competitive sports, for instance, are to Veblen an atavistic survival of the predatory spirit, to Adorno modern sports "seek to restore to the body some of the functions of which the machine has deprived it. But they do so only in order to train men all the more inexorably to serve the machine."[51] Hence the discipline and training in machine culture—whether understood as deprivation (as it is in Adorno's residually "humanist" account) or as adaptation (as it is in Veblen's "technocrat" account)—conserve a movement of regression. The general form this regressive movement takes is (in Habermas's formulation) the "refeudalization of the public sphere" in consumer society.[52]

The modern character of regression includes the relentless melodramas of degeneration and devolution in the discourse of naturalism (charted in the preceding chapter) and the violent cults of atavism and "return to nature" (which I will take up in detail in the final chapter of this study). The connections between the everyday regressions of consumer society and these more dramatic versions of regression (that is, the connections between kitsch and the call of the wild) include also the regressions to romantic property, freedom of contract, and the romances of the market.

The Paper System

The panic about living property (the panic about what Stowe calls "the one great market in bodies and souls") is also the appeal of living property (the perfect unity of persons and things that makes goo "goods," prosthetic extensions of the self-possessed self). The typical American, as we have seen, registers at once this panic and this appeal: the typical American registers the difference between market culture and machine culture in terms of the conflict between self-expression and squaring oneself with a standard. One achievement of the culture of consumption is the conversion of these conflicts into the productive tensions between standardization and self-aggrandizement—the rhythms of suspending and recovering one's self—that operate the sublime economy of consumerism. But in *The American* at least this conflict takes the form of a double discourse, a sort of double-entry bookkeeping, in which one hand writes standards and statistics even as the other continues to write a romance of self-possession and self-identity.

The commercial person, we have seen, admires copies and more than originals; he is fascinated by reproductions and reproduction and by representations and representation. The commercial person's identity, or self-identity, depends on representation. That identity is guaranteed by the imperative of resembling oneself, as a copy repeats its original. But such a dependence of identity on resemblance, such an equation of the self and its representations, must be guaranteed, in turn, by a highly contracted form of resemblance and representation. Self-identity is, for Newman, a "duty" and not a "given," and what this points to is the possibility of a discrepancy and not a simple correspondence between the self and its embodiments, or rather, within the self as embodiment.

Hence the duty of self-resemblance takes on the form of what might be called a *panic of reduction*. Newman everywhere insists on an ideal correspondence between the self and its representations, an insistence nowhere clearer than in his discomfort with the "queer" disconnection between public and private that defines, for the citizen of hotel-civilization and the land of the open door, the problem with the aristocratic form of life. The house that encloses Claire, for instance, presents

> to the outer world a face as impassive and as suggestive of the concentration of privacy within as the blank walls of Eastern seraglios. Newman thought it a queer way for rich people to live; his ideal of grandeur was a splendid facade, diffusing its brilliancy outward too, irradiating hospitality. (50)

What disturbs Newman is the discrepancy between outer surface and interior, the discrepancy between "face" and interior that violates the "physiognomy of goods" in market culture.[53] Similarly, what Newman finds "really infernal" about Mademoiselle Noémie is the contrast between "her appearance" and her acts: "She looked 'lady-like' " (175). Above all, the desire to eliminate the difference between what things look like and what they embody appears in the insistence on the conspicuousness of possessions and hence the aversion to the category of "the secret." "Secrets were, in themselves, hateful things," for Newman: "He felt, himself, that he was an antidote to oppressive secrets; what he offered . . . was, in fact, above all things, a sunny immunity from the need of having any" (151). As Newman simply expresses it, "I don't like mysteries" (195).

"The American" thus desires a natural "fit" between outer face and private interior—a version of *physiognomy*—that effectively coordinates outside and inside, public and private, such that each points to the other. Such an interest in physiognomy, and with the legibility of interior states on the surface of the body that it implies, is frequently made explicit in the realist text.[54] But such a natural legibility appears also in the remarkable accounts that James gives of Newman's relation to language, reading, and writing. "The use that I want to make of your secret," Newman tells Mrs. Bread, is to make it public, or, as Mrs. Bread expresses it, "you want to publish them" (257). The desire to publish the family's secret paper, and to make the family appear in public, reiterates in part Newman's dislike of the inconspicuous and the non-evident. It evokes as well the more fundamental questions that the novel takes up with an extraordinary explicitness and with an extraordinary reductiveness: What does it mean to put things on paper, to write them and to read them? What is the relation between representing and embodying and, collaterally, between reading, writing, and consuming?

The typical American insists on an identity between what things look like and what they mean, and such an identification is most evident, in *The American*, in James's account of Newman's natural, even instinctual, relation to language and meaning. Newman, James observes near the start of the novel, "understood no French," and yet he "apprehended, by a natural instinct, the meaning of the young woman's phrase" (21). Repeatedly, Newman "emerged from dialogues in foreign tongues, of which he had, formally, not understood a word, in full possession of the particular fact he had desired to ascertain" (66). Newman understands "philological processes" in terms of physiological processes: "ascertaining those mysterious correlatives of his familiar American vocables" was "simply a matter of . . . muscular effort on his own part" (24–25). Newman's linguistic materialism equates sounds or material signs with meanings, and if he "formally" does not understand a word, it is the formalist reduction, by which material things resemble the meanings they embody, that defines Newman's philology.

But this natural equation of materiality and meaning, and this fantasy of reduction, are not quite on all fours with the very different understanding of representation the novel also makes explicit. In the earlier discussion of the scene in the museum in which Noémie crosses out her picture, I noted that this scene condenses an opposition between portrait and story, and in effect stages an opposition between representing as drawing and representing as a kind of writing. The crossing out of the picture becomes a legible sign, "the sign of the truth," at once icon and mere line on canvas. The scene thus counterposes two versions of representation: at the one extreme, a radical principle of correspondence (by which things, like portraits, look like what they represent), and, at the other, a principle of discrepancy (by which signs, like writing, represent what they don't look like).

It is not a matter of "choosing" one version or the other. Newman insists on a correspondence between what persons look like and what they are, between outward forms and the meanings they embody, and between forms of writing and what they represent. But Newman, as speculator, relies precisely on the *discrepancies* between "things themselves" and the values or meanings they represent, on the vicissitudes and excesses of value and speculation that, for instance, make possible Newman's astonishingly rapid money-making and self-making. Newman, as consumer, finds fascinating precisely the reproducibility of persons and things. The principle of discrepancy that Newman disavows is thus also the principle he instances: the discrepancy between what things look like and what they represent that for James makes possible stories and persons. The story of "physical capital" in *The American*, as we have seen, involves just such a double account of representation and this twin account is nowhere clearer than in the novel's instancing of what was called, in the later nineteenth century, the American "paper system."

One of the novel's most powerful evocations of this twin account occurs in the scene in which Newman reads the "scrap of white paper" that contains the "secret" of the Bellegarde family. Yet even to call this scene a "scene of reading" is perhaps to proceed a bit too quickly. In this scene of the writing of writing, Newman

> pulled out the paper and quickly unfolded it. It was covered with pencil-marks, which at first, in the feeble light, seemed indistinct. But Newman's fierce curiosity forced a meaning from the tremulous signs. The English of them was as follows. . . .(268)

What is foregrounded in this perverse passage is a hesitating of the act of reading: an emphasis on the recalcitrant physicality of writing and reading; on the unfolding of a scrap of paper; on the indistinctness of the penciled marks on that paper; and on the "translation" of marks into signs and one system of signs into another, a translation that equates construing meaning with the forceful extraction of meaning from "tremulous," as if quasi-animate, signs. The passage thus enacts a minimalist version of the double account of representation I have been describing, bringing into the closest proximity the contradictory insistences of materiality and identity. As in the description, earlier in the novel, of Tristram's walk past the Veronese painting, the American "vaguely looking at it, but much too near to see anything but the grain of the canvas" (26), reading—or looking—in this passage comes close to coming too close to mere marks on paper.

The scene of the letter's reading, like the rehearsed scene of the letter's writing that immediately precedes it, insists on the materiality of the acts of reading and writing, condensing, in shorthand fashion, the physics of representing and reproducing and the becoming-visible of writing as writing. It is therefore not surprising that the letter that Newman looks at, copies, and translates has, as its contents, the "secret" of female violence ("My wife has tried to kill me, and she has done it"), or that the letter is written as if posthumously, with a dead hand ("she . . . put me to death. It is murder"). The novel's consistent linking of the mechanism of the feminine and technologies of copying and reproducing (in the figure of the lethal copyist Noémie, for instance) intimates, in the network of relations we have been tracing, the unnaturalness or machinelikeness of persons.[55] It intimates also the unnaturalness or machinelikeness of forms of reproduction, writing, and representation. Hence the links, powerfully radicalized in turn-of-the-century writing, between writing and the machine, and the links (specified in relation to naturalist discourse in the previous chapter) between technologies of reproduction and death or a machine-like fatality. The reproducibility of writing, like the reproducibility of persons, is the measure of its distance from "nature," or rather, in "naturalist" discourse, the measure of the unnaturalness of nature. As traced in the Introduction and anticipating here the discussions to follow: the anti-naturalism of naturalist discourse and what I have called its realization (in both senses) of "the mechanics of writing" draw into relation the making of persons and technologies of writing and registration. Crucially, this is not because of an essential opposition between persons and machines—as Derrida puts it, "not because we risk death in playing with machines" (playing with a curious piece of machinery like Mme. Nioche, for example). It is rather because the capacity to represent and reproduce that, by this account, makes up persons also makes visible the technology of writing. And such a becoming visible of the technology of writing in machine culture risks making visible the links between the materiality of writing and the making of persons, and thus the internal relations between persons and machines.[56]

There is a good deal more to be said about moments such as these, and I will be returning to the radicalization of the nexus of writing and mechanics in machine culture in the next part of this study. Here I want to suggest instead the process by which James's text monitors and allays such recognitions.[57] One way of allaying or monitoring such a linking of persons and machines is by way of its "localization" (for example, in the scandalized representation of the mechanism of the feminine, that is, the female *as* machine). Another is by way of what I have been calling a panic of reduction. It is precisely such a panic of reduction that, as we have seen, enables Newman's "translation" of materiality into identity, and *The American* progressively tests out what this forcing of meaning looks like. Such a reduction or minimalism characterizes the final pages of *The American*: the narrative turns more and more to paper itself. I am referring only in part to the obsessive proliferation of images of papers, writing, and reading toward the end of the novel. One finds here not merely an emphasis on the "marquis's manuscript" (267) as the "little paper" (274), as "a scrap of white paper" (268), as "the paper—the paper" (265); nor merely a reiterated concern with reading and writing ("he read it. He had more than time to read it" [283]; "Wait till she reads the paper!" [258]). One finds as well a turning of persons, acts, and bodies to paper: "Newman made a movement as if he were turning over the page of a novel" (262); "her thin lips curved like scorched paper" (281); "he turned to the marquis, who was terribly white—whiter than Newman had ever seen anyone out of a picture. 'A paper. . .' " (282). One finds, finally, a minimalism that equates the consuming of paper and of the novel itself. Even as Newman's burning of the little scrap of white paper is a way to "close the book and put it away" (306), the representation of the paper's consumption and the reader's consumption of James's representation are made to collate perfectly in the novel's final sentence: "Newman instinctively turned to see if the little paper was in fact consumed; but there was nothing left of it" (309). The slight hesitation ("but")—marking the desire to see, or fear of seeing, nothing left to see—plangently reinforces this closing of the book. But such a psychologizing barely reanimates this becoming visible of the materiality of representation and, correlatively, of the artifactuality of persons.

The novel thus ends by drawing together the physical act of consuming and the matter of representation. The turning of persons to paper forms part of the same set of relations. The novel's insistent return to the matter of representation indicates not simply a certain narrative self-reflexivity but the entanglement of the "topics" of reproduction and representation with the hybrid forms of "physical capital," "living property," and "commercial persons." Collaterally, Newman's contractual identity is inseparable from paper forms. To promise to leave Europe is to "sign a paper never to come back" (219); in his attempted "bargain" with the Bellegardes, "a simple *yes* or *no* on paper will do" (253); to enter into relations for Newman is to sign a contract, "I never sign a paper without reading it first" (201). Newman's "business papers" (301) and the little piece of white paper, containing the family secrets, that he tries to exchange for Claire, are instances of the "paper system" that dominated later nineteenth-century interests both in money and in representation.

The two dominant national debates during the "greenback era" (James's novel opens in 1868, appears in 1875–76) are debates about the peculiar institutions of living property: the debate about "reconstruction" and property in persons, and the debate about the "paper system" and the embodiment of value in representations. If precious and "hard" metals were also precious symbols of a medium of representation that looked like what it represented, and if paper money disturbingly represented what it didn't look like or feel like, these rival tendencies were also implicit in the double-entry system of "the gold standard" itself (gold as value-intrinsic, the standard as value-systemic). It has been argued, about the effects of the "paper system," that "the apparently 'diabolical' 'interplay of money and mere writing to a point where the two be[come] confused' involves a general ideological development: the tendency of paper money to distort our 'natural' understanding of the relationship between symbols and thing."[58]

But it would seem rather that what is at work in "mere writing" like *The American* is not so much the distortion of "natural" symbols by paper representations as it is the systematic "floating" of these different modes of value and representation that defines the system of physical capital. George Eliot observes of the physician Lydgate in *Middlemarch* (1871–72) that, in his youth, "he had no more thought of representing to himself how his blood circulated than how paper served instead of gold."[59] It's this circulation of self-representations, representations of the body, and economic representation that makes up the market in persons and the physiognomy of physical capital.

Market Culture and Machine Culture

The great difference between a person and a thing is, I have argued, one of the centering subjects of the American novel in the third quarter of the nineteenth century, a period conveniently if roughly marked, at one end, by Stowe's investigation of America's "peculiar institution" of slavery—"one great market in bodies and souls"—and, at the other, by James's account of the "peculiar institutions of his native land" (289). What defines these institutions are the shifting relations of living property. I want to close this investigation of the tensions between the forms of market culture and machine culture by taking up these relations from a somewhat different point of view and by indicating some of the ways in which these tensions are conserved in recent accounts of "the American" and in some recent accounts of American culture generally.

In his preface to *The American*, written thirty years after the novel's first publication, James retroactively clarifies the tensions between market culture and machine culture that the novel instances. The famous, albeit famously elusive, distinction between "romance" and "realism" set out in the preface becomes legible precisely in terms of the counter-tendencies of market culture and machine culture. The difference between romance and realism is figured in terms of the difference between the "romantic property of my subject" that James aligns with the market, and the cases, measures, and standards that James aligns with the realisms of the machine process.

On the one side, James in the preface links his own "labour" of writing to the labor of embodiment and the process of physical reproduction. This is a process of "germination" and gestation that makes "my conception concrete" through a "notable increase in weight" and an eventual "filling itself with light in that air" (1, 4, 3). But what above all inflects this account of conception is not merely the familiar connection between composition and gestation but, more precisely, the connection between conceiving or making persons and owning them. The "happy development" of the subject—the way "things grow up and are formed"—is also an intensity of possession and the possession of another person (or of oneself) as living property: "the intensity of the creative effort to get into the skin of the creature; the act of personal possession of one being by another at its completest" (7, 13). The romantic property of the subject depends on the market in persons. And the Jamesian "sublime economy" depends on the circular relation between persons and possessions, and between affective and economic conditions, that makes up the romance of the market. Thus, the insistent punning into equivalence of affective and economic states in the Jamesian idiom (the aesthetic "economy" of "interest," "value," and "appreciation," for instance); thus, the familiarly Jamesian syntax of perfect balance and reciprocity, the "beautiful circuit . . . of our thought and our desire" that makes up the circular or tautological desire-economy of the romance (9); thus, the relays between the form of the romance and the logic of market culture.

What James sets in opposition to the romantic property of the subject is the "indignity of a sliding scale and a shifting measure" (10). This is the indignity of squaring oneself with a standard, with variable scales, sizes, measures—with, "in short, as they say of collars and gloves and shoes, the size and 'number' " and with, in short, the "class" and "classification" that "fits all its cases" (10). This is, moreover, the opposition between the "absolute" standards of good and evil that provides, for James, the "thrill" of the romance, and the "sliding" scales and standards of the culture of mass consumption, the sizes and styles and numbers of collars and gloves and shoes.

If the process of composition is, on the one hand, a matter "just of his feeling and seeing, of his conceiving, in a word, and of his thereby inevitably expressing himself" (8–9), it is, on the other, a matter of "plotting and planning and calculating." The writer, writing as if one hand did not know what the other was doing, makes double entries and "commits himself in both directions" (9). If the first is a matter of conception and "inevitable" self-expression, the second is a matter of auto-taylorization. The work of plotting, and planning, and calculating appears in terms of the author's "daily effort not to waste time" and in the adjustment to the time-table of "the economy of serialization" (1); in the "complex of fine measurements" by which "anything is producible" (8); and in the generic classifications by "quantity and number" (9) that pressure this preface on cases, types, and generic classifications (the cases and classes of "romance" and "realism," for example).

The difference between romance and realism thus appears in terms of (and protects the terms of) the difference between the desires of the market and the disciplines of mass culture: the discipline of squaring oneself with "the inconvenience of a *related*, a measurable state, a state subject to all our vulgar communities" (10). The vulgar communities of these measurable and united states are set in opposition to the intimacies of personal possession: "our pleasure . . . in these intimate appreciations (for which, as I am well aware, ninety-nine readers in a hundred have no use whatever)" (8). But the invocation of the idiom of the consumer survey and marketing sample (99 in a 100) registers in miniature an extraordinary transition in what it means to be "representative." The shifting measures of persons and things redefine typicality and the meaning of the standard, such that the statistical average and norm serve to correlate measure and humanity, that is, measure and individuality. The correlation of persons and measures makes possible the cases and classes and sizes that, in turn, individualize individuals. This is what makes it possible for the bell-shaped curve of statistical regularities and deviations to serve as a fundamental index of identity and individuality: not the individual universal (the "type" of Lukácsian realism) but the statistical person (the American as the typical).

Such a transition in what it means to be representative is perhaps most pressing in the redefinition of what it means to "account for" persons and their actions. The tensions between what might be called systemic and free-market or laissez-faire accounts of persons and actions are not hard to locate. As Emerson noted early on, in his essay "Fate," "the new science of Statistics" (which he discusses along with the new "sciences" of physiognomy and heredity) makes visible "the stealthy power of other laws which act on us daily . . . organization tyrannizing over character." One implication of such a daily and ordinary statistical fatality—instanced for Emerson by the social numbers of Adolphe Quetelet, "the great regularity salesman of the nineteenth century"[60]—is the detection of "a rule that the most casual and extraordinary events . . . become matter of fixed calculation."[61] But if for Emerson the "tyranny" of statistical "organization" over "character" is the tyranny of being acted on rather than acting, for James's American, there appears to be, at least at certain moments, not an opposition between statistics and doing but rather an identification of statistics and doing: Newman, James notes, "was fond of statistics; he liked to know how things were done" (55).

In *The American* the unsteady transition in what it means to be typical or representative, and the correlative problems of accounting for persons and actions, appear, as we have seen, in the general terms of a perplexity about "whether to attribute" expressions and acts "to habit or to intention, to art or to nature" (115). The uncertainty about the naturalness or artifactuality of persons and acts appears in terms of the "timeless" opposition between individual will and impersonal force. Experiencing setbacks in his attempts to "wrest a fortune" from "impertinent force," Newman momentarily confronts a "mysterious something": "there seemed to him something in life stronger than his own will" (31–32).

As Ian Hacking has recently observed, describing the conflicts between statistical and anti-statistical understandings of persons and acts, "normality is like determinism, both timeless and dated, an idea that in some sense has been with us always, but which can in a moment adopt a completely new form of life."[62] One of the new forms of life that the determinism/will polarity takes, in the later nineteenth century, is the statistical/anti-statistical polarity. And one form the statistical/anti-statistical polarity takes is the conflict between the logic of market culture and possessive individualism and the logic of machine culture and disciplinary individualism. Put as schematically as possible, whereas the laissez-faire or atomistic conception of the individual protects an essential opposition between persons and cultures, the systemic conception of the individual understands persons as an effect or reflex of their culture. By this "collective" statistical understanding, "it was not the case that individual people were constrained in their freedom by belonging to a culture. For they had no atomic individualistic self to be constrained, until they were human beings within a culture."[63]

These alternative accounts of the individual are familiar enough, and not least because these alternatives have continued to structure a wide range of cultural criticism. From one point of view, it might be argued that there is simply nothing very "deep" to be said about the individual vs. culture, agency vs. structure, act vs. system problem. That is, if we understand agency and intention not as the *cause* of an action but as *part* of an action, then both the understanding of the individual as constrained by her or his culture and the understanding of the individual as a reflex of her or his culture seem misleading. By this view, there is simply nothing interesting in general (nothing both interesting and in general) to be said about matters of agency and intention or about the relations of individuals and culture, although there is, therefore, a good deal interesting in particular to be said about these matters.[64] But such a "mixed," immanent, and avowedly self-divided notion of agency and intention as part of an action rather than as the cause of an action clearly has seemed less acceptable or less appealing than the all-or-nothing accounts of agency and agents that continue to dominate cultural criticism. That criticism continues to be dominated by the "sublime" melodrama of uncertain agency and by rehearsals of the melodrama of agency-suspension and agency-recovery. What I have been suggesting is that such a melodrama closely resembles the rituals of self-possession and self-identity that make up the logic of market culture. It is not merely that the antinomies of the agency/system polarity restate the "liberal" antinomies of the market but also that the abstraction and generalization of such antinomies tend to abstract and generalize a fantasy of the market, a fantasy of possession and self-possession.[65] If a range of recent cultural history is insistently drawn to large abstractions of "the market" and "exchange" and "negotiation," and to large metaphors of "property" and "possession," this is at least in part because "the market" functions in that work at once as topic and as metaphor. Or rather, the logic of equivalence that is taken to define the market makes metaphorics and the market two ways of saying the same thing. By this logic, the very notion of a "cultural poetics" is already a tautological one.[66]

What this "regression" to a fantasy of the market amounts to can be further specified by considering briefly some recent rehearsals of market culture. In the investigation of capitalism and interests I've already drawn on, Thomas L. Haskell provides a significant critique of what he calls the "social control thesis." This is, in short, the explanatory thesis that regards beliefs and acts as ultimately an index or reflex of economic or class interests (what we might call, on the model of the genetic fallacy, the *interests fallacy*). By this thesis, acts and beliefs flow from economic interests, and the effects of acts and beliefs are understood as the product of economic interests, whether these effects are consciously intended (the work of a visible hand) or self-deceptively or unconsciously intended (the work of an invisible one). For Haskell, the problem with the social control thesis is that it attributes a "greater degree of intention . . . than the evidence can substantiate." That is, this approach closes the gap between intentions and consequences, and between subjects and structures, but only by converting *unintended* consequences into *unconsciously intended* consequences.[67] "People know what they do," Michel Foucault remarked, and "they frequently know why they do what they do; but what they don't know is what what they do does." In his *Outline of a Theory of Practice*, the sociologist Pierre Bourdieu observes: "It is because subjects do not, strictly speaking, know what they are doing that what they do has more meaning than what they know." Whereas Foucault's remark indicates an ineradicable gap between intentions and consequences, acts and meanings, Bourdieu's bridges this gap, but only by regarding apparently unintended consequences as unknowingly or unconsciously intended consequences. These two statements together could be taken to epitomize the rival accounts of the subject/structure problem we are considering.[68]

The solution, for Haskell, among others, is to hold on to a version of economic determinism while eliminating the tie between economic structures and the intentions or interests of persons or classes. As he expresses it, the "rival scheme of explanation I advocate retains the claim that there is a 'process of determination,' but deliberately abandons the claim of intentionality." It quickly becomes clear, however, that the version of determination Haskell retains represents not an abandoning of, but in fact an even greater degree of, intentionality. Haskell understands the rise of capitalism and what he describes as the origins of the humanitarian sensibility in terms of "the forms of life the market encouraged," and throughout he emphasizes the "autonomous power of the market to shape character." What such a notion of the market's power involves, however, is precisely an attribution of intentionality to economic structures themselves. One might say that for Haskell the problem with the social control thesis is not that it attributes too much intention but too little: if the market shapes the character of the subject of capitalism, this is because the market has itself become a subject, an agent with intentions, interests, and powers.[69]

Such a personifying of the market has of course a long history: in fact, I have been suggesting that the logic of the market and of personification (living property) are scarcely separable. If on this account the dominance of the market makes persons into things, it also makes things, like tables or chairs or markets, into persons. As Thorstein Veblen observes, in a chapter of *The Theory of the Leisure Class* called "The Belief in Luck," the "animistic sense of relations and things, that imputes a quasi-personal character to facts" survives side-by-side with an "industrial organization [that] assumes more and more of the character of a mechanism"; the animistic "habit of mind" survives, for instance, in the "modern reminiscence of the belief . . . in the guidance of an unseen hand."[70]

Which is to indicate the modern reminiscences of the market habit of mind—the fantasy of market culture—within the "industrial organization" and "mechanisms" of machine culture. Stated very generally, recent accounts of the opposition of subject and structure have tended to reproduce, in the form of a theoretical neutralization, the liberal antinomies of market culture: the tensions between things and persons, mechanism and animism, materiality and meaning, the body and embodiment.[71] Not unexpectedly, the still-governing opposition between the "constituent subject" and the "economic in the final instance" itself appears as a version of the twin determinations of the market: that is, if the market requires the "sovereign individual," the regularities in exchange supported by the dominance of the transpersonal system of the "self-regulating market" and its "laws" were conducive not merely to belief in determinism but also to a logic of infinite exchange and interconvertibility.[72]

These recent accounts have thus tended toward a theoretical duplication of this contradictory logic. For Haskell, for instance, the market imposes "forms of life" and the point is to abandon claims of intention and to trace these forms, "the subtle isomorphisms and homologies that arise from a cognitive style common to economic affairs."[73] The "new historicist" Walter Benn Michaels, in a richly provocative study of late nineteenth-century economic-literary relations, traces the ways in which an imperative of personification "constitutes the possibility of bourgeois economy." But in the new historicist account of the market, as in Haskell's, "the whole point of [the] analysis . . . is, by subverting the primacy of the subject in literary history to subvert also the primacy of interest"; the aim is thus to locate "positions and their negations" in a "double logic" and "in so doing to suggest one way of shifting the focus of literary history from the individual text and author to structures whose coherence, interest, and effect may be greater than that of either author or text."[74] Yet since this very opposition of subject or individual and structure defines the "form of life" or "double logic" of market culture, the question becomes: How do these moves from subject to structure avoid a repetition and *formalization* of the logic of the market, avoid what in effect amounts to a purely economistic or logicist positing of isomorphisms and reversible or fungible oppositions? How, in short, do they avoid a rehearsal of the sublime economy of the market itself?[75]

The effect of such a formalization of the logic of the market is not hard to detect. The new historicist account of the market is not historical but methodological or theoretical: that account *posits* the circular relation between individual desires and social demands (the theoretical equivalent of a sort of taylorization in consumption) that backs the intricated notions of possessive individualism and market culture. But the anachronistic description of turn-of-the-century American culture in terms of the logic of the market not merely makes for basically inaccurate accounts of the machine culture of consumption. As C.B. Macpherson has suggested, the disarticulation of economic relations and political obligations, and the rise of modern social classes in the space this tension between economic and political forms opens, had, at least by the middle of the nineteenth century, challenged the political theory of possessive individualism and the logic of market culture: although "possessive market relations continued to prevail," the "development of the market system, producing a class which could envisage alternatives to the system, thus destroyed the social fact (acceptance of inevitability of market relations) which had fulfilled the first prerequisite of an autonomous theory of political obligation."[76] One effect of the new historicist account of the market is precisely to reduce these tensions between political and economic relations to a general logic of equivalence; another is to posit exactly the inevitability of market relations, in effect *as* the circular logic of culturalism *as such*. The appeals of that logic, and its limits, should by now be clear enough.

I have been arguing here that turn-of-the-century American consumer society operates by attempting to solicit and to manage the rival tendencies of possessive individualism and market culture, on the one side, disciplinary individualism and machine culture, on the other. Such a deployment of difference involves in part the perpetual adjudication of what Foucault has called "the contractual link" and "the disciplinary link." What this entails is the sustained interplay between the formal and egalitarian principles and rights, underwritten by the theory and logic of contract (the formal framework of market culture) and the "everyday, physical mechanisms" of discipline, normalization, and classification against which, but also through which, these formal, juridical principles are brought into play. As Foucault summarizes it: "The general juridical form that guaranteed a system of rights that were egalitarian in principle was supported by these tiny, everyday, physical mechanisms, by all those systems of micro-power that are essentially non-egalitarian and asymmetrical that we call the disciplines":

> The real, corporal disciplines constituted the foundation of the formal juridical liberties. The contract may have been regarded as the ideal foundation of law and political power; panopticism constituted the technique, universally widespread, of coercion. It continued to work in depth on the juridical structures of society, in order to make the effective mechanisms of power function in opposition to the formal framework that it had acquired. The "Enlightenment," which discovered the liberties, also invented the disciplines.[77]

I have been suggesting some of the ways in which the culture of consumption refunctions and radicalizes these oppositions and tensions. I have been arguing, for example, that it is such a "flotation" of the formal equivalences of market culture (the contractual relay) and the bodily disciplines, measures, and standards of machine culture (the disciplinary relay) that makes for the rival determinations of the individuality of the individual in consumer society. And I have been tracing the melodramas of agency, and the rituals of possession and self-possession, that sustain the anxious renegotiations of the problems of the body and of personhood, and the particular forms these renegotiations take, at the turn of the century and beyond.

I will be reexamining these problems and forms, in greater detail and to somewhat different ends, in the chapters that follow. But I want to close this account of the logic of the market by returning for a moment to *The American* and to several passages that epitomize at once the appeals and panics of market culture centered on here. In "The Naturalist Machine," I argued that one alternative to the opposition of an empty voluntarism and an anonymous social mechanism might take the form of a "biopolitical analysis": an investigation of the disposition of the subject at the point of intersection of bodily and cultural forms and practices, what amounts to a cultural logistics. This biopolitics fashions what Pierre Bourdieu has called the "socially informed body," the embodiment both of the dispositions of bodies and the exchanges of capital and meaning. Such a coordination of the social body operates in the most trivial and everyday practices and techniques, in all those small adjustments, at once public and private, that exact the essential while apparently requiring the insignificant. One form the fashioning of the socially informed body takes is the production of a specific cultural physiognomy: for instance, the physiognomy of the market that appears in *The American*.

Such a physiognomy of the market can be seen in the tautological reduction of persons and conditions in market culture. It is visible, for example, on the opening page of *The American*, in Newman's "profound enjoyment of his posture," "head thrown back and legs outstretched," that denotes the "serene possession" of the commercial person: "his physiognomy would have sufficiently indicated that he was a shrewd and capable fellow" (17). Or, to take another example, from the following page of this story, a story in which looking is everywhere linked, as we have seen, to relations of possession: "it was our friend's eye that chiefly told his story." The paradoxical look readable in that eye constitutes, for James as for Marx, the paradoxical disposition of the capitalist: "It was full of contradictory suggestions . . . Frigid yet friendly, frank and yet cautious, shrewd yet credulous, positive yet skeptical, confident yet shy, extremely intelligent and extremely good-humoured, there was something vaguely defiant in its concessions, and something reassuring in its reserve" (18–19).[78] The very excessiveness of James's "reading" of the businessman's eye foregrounds the extraordinary legibility of this socially informed body. And what becomes legible here is not so much the identity of the individual as it is the paradoxes, internal divisions, and tensions that are "coordinated" in the fashioning of that identity.

On this view, the tensions between materiality and identity, and between self and structure, we have been considering appear somewhat differently. If the realist novel relies on such a social physiognomy, the novel at the same time, we have seen, insists on the discrepancies between the subject's will and the structure he inhabits, insists on a certain disarticulation of private dispositions and "something in life stronger than his own will." It may by now be possible to understand this contradiction as something more than an instance of realism's "double logic" or as an embodiment of the contradictory determinations of the contractual self. These contradictions and differences are adjusted and coordinated in the novel: at once solicited by the novel and tactically deployed and managed. And this solicitation and management of a logic of difference reveals the logistics of market culture and the forms of representation and forms of reduction on which it depends.

The duty of the subject of realism is to fashion a character that corresponds to its representations (the duty of resembling oneself). The duty of later nineteenth-century realism is to fashion a subject who corresponds to, exemplifies, and embodies the cultural meanings that, completing the tautology, inform him. This is the cultural work immanent in the realist imperative of the social legibility of character: it is, on one level, the constant adjustment of subject and structure, and, on another, the perpetual coordination of bodies and meanings, content and form, that the realist novel enacts. The realist criteria of legible social types, consistency of character, and deterministic and linear narrative progress function to secure at once the intelligibility and the supervision of character and event.[79]

The realist text, however, is not merely *about* the fashioning of socially informed bodies; the novel also *underwrites* such a fashioning. The socially informed body of the character of realism may appear as a second nature, as, for instance, the "physical capital which the owner does nothing to 'keep up.' " The achievement of such an "almost ideal completeness" would seem to be something of an advertisement for an ideally complete realism. But the realist body is never quite so well-informed, never quite so autonomously complete and unrelated: in the serialized form of the installment plan, it is always "to be completed."

For the realist novel there is no end to the work of "keeping up"; no end to the work of disclosing and accounting for the discrepancies between what things and persons look like and what they "are"; no end to the work of "adjusting" subjects and structures, coordinating the everyday practices, disciplines, and dispositions of the body and the formal and juridical principles in relation to and against which these practices function. And if these practices are never quite faithful to these principles, the realist text capitalizes on this discrepancy, advertising the differences between persons and things, will and structure, content and form even as it extends the relays between them. The making of Americans at the turn of the century, and the rituals of consumer culture, are realized through the tensions between market culture and machine culture and between competitive and disciplinary determinations of the individual. There is always physical capital to be made, and made, as both writers and businessmen in the American paper system are well aware, out of little pieces of paper. The consumer of the realist text reenacts—that is, embodies—these adjustments and these tensions as with the turning of a page.

Part III

Statistical Persons

How the Other Half Looks

There is a small scene of theater in Stephen Crane's *Maggie: A Girl of the Streets* (1893) that concisely invokes one of the dominant projects of later nineteenth-century realism: what might be described as the project of "accounting for" persons. "A ventriloquist followed the dancer," Crane writes: "He held two fantastic dolls on his knees. He made them sing mournful ditties and say funny things about geography and Ireland.'Do dose little men talk?' asked Maggie. 'Naw,' said Pete, 'it's some damn fake. See?' "[1] What the barely articulate Maggie begins to see in seeing the difference between talking dolls and persons, between imitation persons and real ones, is at once the possibility of imitation itself and the dependence of persons on the possibility of representation and imitation. This is made a bit clearer in Crane's commentary on the effects of a more extended scene of theatrical representation a little later on in the novel: "The theater," Crane observes, "made [Maggie] think. She wondered if the culture and refinement she had seen imitated, perhaps grotesquely, by a heroine on the stage, could be acquired by a girl who lived in a tenement house and worked in a shirt factory" (28). We might say that Maggie begins to think and to wonder—indeed, in the terms of the story, gains an interiority or becomes a person—by internalizing a desire to imitate that is also a desire to transcend at once her material conditions and her inarticulate, material condition. We might say further that Maggie's mistake in personifying the fantastic dolls is the counterpart of her desire to acquire the character she has seen imitated: in effect, a desire to personify or to *impersonate* herself.

One way of accounting for this desire involves the social positionings these scenes enact. The play that makes Maggie think, for instance, posits an "inevitable" difference between those who are socially included and those who are excluded. "Maggie lost herself in sympathy with the wanderers swooning in snow storms beneath happy-hued church windows. And a choir within singing 'Joy to the World.' To Maggie and the rest of the audience, this was transcendental realism. Joy always within, and they, like the actor, inevitably without" (27). By the logic of what Crane calls "transcendental realism," if the play at once ratifies an inevitable difference between those who are included and those who are excluded, it also and paradoxically promotes a desire to transcend this difference and, as we have seen, to imitate the privileged interior (that is, to reform along the lines of middle-class values). By this paradoxical logic, the representation itself effects at once a fantasy of reform and an "inevitable" containment, effects a sort of cultural policing and self-policing of the underclasses. But there is perhaps something more involved here than the disciplinary effects of slum theater, something more because, in Crane's account, the desire to occupy the privileged interior appears also as the desire to *have* an interior: being inside and having an inside are the two sides of a single formation here. The subject of realism is formed from the outside in—filled, as it were, with the social—and the project of accounting for persons in the realist text becomes then an account of persons as "socially constituted."

The radical pressure that the later nineteenth-century realist and naturalist text places on the category of persons—in positing that the individual is something that can be made—is bound up with an imperative that this making refer back to the constitutive agency of the social.[2] And if such an account has become a virtual "given" in some recent versions of social criticism and new historicism, such continuations of "the realist project" have perhaps duplicated as well the counter-side of this given. My concern in these pages then is both with some of the recalcitrant forms the realist and naturalist account of the subject takes and with the implications of these forms for more recent cultural criticism.

Put simply, the realist imperative involves not merely the imposition of a specific political and social program, but also and more generally the programmatic insistence that "everything is political" or social "all the way down." This imperative is, however, at once opposed to and coupled with a very different way of representing agents and agency. Mark Twain's *Connecticut Yankee* (1889) provides perhaps the most schematic instance of the double discourse, the twin determination of the individuality of the individual (persons as "naturalist machines"), that we have been tracing in machine culture.[3] It's not hard to see that the narrative insists at once on the artificial and on the natural character of the individual. On the one side, "training, training is everything," and Hank Morgan's Man-Factories "turn groping and grubbing automata into men" (89); but, on the other side, and set against this Taylorite disciplinary scenario, there is the notion that "a man is a man at bottom" (172) or, most basically, that "there is no accounting for human beings" (113).

In the late nineteenth-century novel, as we have seen, these two ways of representing and accounting for persons and their natural or social or artifactual constitution are most insistently figured in terms of the body and the machine. I have argued, in the preceding parts of this study (and it is this argument that I want to extend and redirect here), that the late nineteenth-century realist and naturalist text operates through a double discourse by which the apparently opposed registers of the body and the machine are coordinated and "managed" within a flexible control-technology of regulation and production. We might recall here Thorstein Veblen's counterposing of what he invidiously calls "the radiant body" against "dispassionate," "mechanistic sequences of cause and effect" (see Part II); or Frederick Winslow Taylor's disarticulation of natural bodies, mapping of the body onto the machine, and reembodiment of individuals as the "head" and "hands" of a new corporate whole (see Part V); or the naturalist writers' devising of the "naturalist machine" as a counter-model of the generation of persons (see Part I). The relays elaborated between bodies and machines in this discourse hold steadily visible the problem of accounting for human beings.

"Hold steadily visible" at least in part because this body-machine complex cannot be considered apart from the "realist" insistence on a compulsory and compulsive visibility.[4] The frequent associations of later nineteenth-century realism with a sort of dissection, vivisection, or surgical opening of the body are well known. But the association of realism with these technologies of the body points also to the realist imperative of making everything, including interior states, visible, legible, and governable. As W.C. Brownell observed in the *Nation* in 1882 (in the context of a review of Henry James's *The Portrait of a Lady*), the characters of the realist novel are "turned inside out for the reader's inspection."

This imperative involves at once a fantasy of surveillance and a requirement of embodiment. That is, the realist desire to see is also necessarily a desire to make visible: to embody, physically or materially, character, persons, and inner states and, collaterally, to "open" these states to what Crane calls the "machines of perception," what

Zola calls "the mechanism of the eye," and what the turn-of-the-century reformer, police reporter, and photographer Jacob Riis calls the social technologies of an "eternal vigilance." From one point of view, the realist project of making-visible is perfectly in line with the techniques of a certain social discipline, along the lines that Foucault has mapped in detail: the opening of the everyday ordinariness of every body to, and the fabrication of individuals under, the perfect eye of something like the police. But, from another perspective, the requirement of embodiment, of turning the body inside out for inspection, takes on a virtually *obstetrical* form in realist discourse. If the first takes as its model the man-factory and the mechanical reproduction of individuals, the second, we have seen, takes as its model the figure of the mother and the biological making of persons. What the logistics of seeing and embodiment in the realist text entails, then, is a perpetual negotiation between these two models of personation: between the body (specifically, the more radically embodied and embodying maternal body) and the technologies of the social machine.

Such a negotiation between these opposed but coupled models structures, for instance, both Crane's *Maggie* and Riis's *How the Other Half Lives* (1890), and, before turning to some more recent versions of this logistics, it may be useful to indicate in a bit more detail how Crane's and Riis's accounts of the other half look. Relations of power in the realist text are insistently articulated along lines of sight. More specifically, the realist vision of the urban underworld posits and fantasizes a disciplinary relation between seeing (seeing and being seen) and the exercising of power: the realist investment in seeing entails a policing of the real. It entails also the complex and tense interaction between vision and embodiment, between *the visual* and *the corporeal*, we have begun to sketch. This tense interaction makes for the excitations generated by relations of vision and supervision: one discovers in the realist fascination with seeing, and not least in the spectacles of violence, and thrilled identification with representations, in the realist text, an eroticizing of power and of the power of making-visible. And this eroticizing of power is nowhere clearer than in the almost programmatic rewriting of the story of the (social) "other half" in terms of the story of the (sexual) "lower half": more specifically, in terms of the counterpart stories of the fallen girl and monstrously prolific mother of the slums.

Crane's stories virtually inventory relations of seeing, vigilance, and power. But the desire to see that marks Crane's stories, the aesthetic and erotic fascination with a vision that is also a supervision, always also implicitly calls forth its opposite: the dangers and vulnerabilities inherent in that desire. In Crane, as in Zola, the struggle to see involves not merely the (Foucauldian) scenario of a disciplinary relation between seeing and exercising power but also the possibility of a dangerously corporeal absorption in a theatricalized watching. This interplay between "absorption and theatricality" structures and "operates" the realist fantasy of surveillance.[5] The violent confrontation between Maggie's brother Jimmie and her seducer Pete, for instance, produces, in the crowd that quickly gathers to view the spectacle, "an absorbing anxiety to see" (37). Again and again in *Maggie*, a "wide respectable circle of interested watchers" (28) gathers about scenes of slum violence.

Yet if the crowd's anxiety to see immediately summons the police to the scene, what is at work here is not exactly an equation of watching and policing nor even the related conversion of slum violence into the theatricalized entertainment of the "respectable," although the scene's logic involves both of these. What complicates the policeman's "interruption" is a temporary loss of "balance" in these relations of seeing and power, an imbalance produced by the excited engagement or absorption of both policeman and crowd. The officer of the law quickly "regain[s] his balance" (a steadiness at once physical and official) and resumes a position of "disengaged" power, converting violence to spectacle: "Well, well, you are a pair of pictures." Yet through a further turn in the balance of power, it is "the crowd-encompassed policeman and his charge" themselves who, in turn, finally engage the interest of "the more law-loving, or excited individuals of the crowd" (37–39). Policing itself has become an absorbing and entertaining spectacle (as in the detective fiction of the 1890s). But the love of law appears here, Crane's ambiguous phrasing suggests, as equivalent to, or even as a mere by-product of, an excited love of seeing. (Law-loving "or" excitement seem uncertainly alternatives and equivalents.) And these excitations cannot be separated from the risks and vulnerabilities of a deep engagement at once visual and bodily. Above all, Crane insists on the ways in which the power of seeing is quickly disrupted by the pleasures of seeing: the very absorption in, even intoxication with, seeing opens the possibility of violent loss of balance or *dis*empowerment. This is a seeing machine in which everyone is caught, hesitating both equations of looking and power and the looker's privilege of relative disembodiment (hesitating, that is, the difference between the disembodied gaze and the corporeality of the eye).

It would be possible to trace the operations of this realist seeing machine, and its shifting relations of control, in much greater detail.[6] If, for instance, the fascination with seeing opens the possibility of becoming vulnerably "transfixed with interest" (52), it is therefore the character's *resistance* to seeing that seems to provide a way of neutralizing such fixations. One form this resistance takes, in Crane's *Maggie*, is the cultivation of what Crane calls "the self-contained look" (24): the turning of a paradoxically unseeing "glare upon all things" (14), or the intransigence of "returning stony stare for stony stare" ("Dat jay has got glass eyes"), or the trance-like seeing that is not finally a seeing at all (26). Yet it is not hard to detect how the "self-contained" look, in these cases at least, effectively reinforces the radical self-containment of the closed world of the slums. Jimmie is only "awaken[ed]" from "a sort of trance of observation" by the violence of "some blue policeman turned red" (15). The self-contained look might be described as a *recidivist* way of seeing, as a way of seeing that makes the entranced observer the agent of his own containment. Hence the recidivist way of seeing ratifies the closed circuit between self-containment and the police, constitutes what Foucault has called a "self-absorbed delinquency."[7]

What complicates this "resistance" even further are the very fascinations produced, in turn, by these trance-like states. Just such a self-containment, just such an "enticing nonchalance" (17), appears as powerfully attractive or charismatic throughout the novel. What makes the slum world fascinating to look at, both for watchers and for readers of a self-contained slum violence, is at least in part its charismatic self-absorption: the irresistible attraction of a certain narcissism. "It seems very evident," Freud observes, "that another person's narcissism has a great attraction for those who have renounced part of their own narcissism and are in search of object-love. The charm of a child lies to a great extent in his narcissism, his self-contentment and inaccessibility, just as does the charm of certain animals which seem not to concern themselves about us, such as cats and the large beasts of prey."[8] The realist fascination with seeing, and the identification with representations described in and produced by the realist text, disclose what Naomi Schor has, in a different context, characterized as an "aesthetics of narcissism."[9] It is worth recalling here Crane's obsessive return to stories of autarchic, small children (for instance, "An Ominous Baby" and "Death and the Child"), animals (for instance, "A Dark-Brown Dog" and "The Snake"), and, above all, to the bestial and autogenic violence of the self-contained slums. But these two very different but closely linked versions of the self-absorbed look, as recidivist and as charismatic, are most concisely embodied in the twinned figures of the prostitute or "girl of the streets" and the monstrous mother of the slums.

If the realist and naturalist novel frequently seems to require the figure of the prostitute, this is because the case of the fallen girl provides a way of at once embodying and bringing to book, in both senses, the desire to see and the project of making "the social" visible. For one thing, the compulsory visibility of the prostitute—the painted woman, who must catch one's eye—draws on (as Zola's *Nana* makes clear enough) an exciting theatricality or illusionism and a charismatic self-absorption and self-abstraction. As Crane represents it, in a passage that instances, with an astonishing condensation, this erotic trick of the eye, the "good looker's" thrilled interplay of the visible and the bodily: "Painted with no apparent paint," Crane writes, "she looked clear-eyed through the stares of men" (43).

For another, the realist seeing machine disposes what Crane describes as "the machinery of justice" that quarantines the controlled illegality of prostitution.[10] Beyond that, the figure of the fallen girl embodies not merely the realist interest in vision and supervision, but also, and with a certain irony, the realist project of accounting for persons as "socially constituted," that is, as social artifacts. The single figure projected by Crane and Jacob Riis, among others, as escaping this way of accounting for persons is the utopian and transcendental figure of the *un*fallen girl. This is therefore not a material girl but something like a virgin of the slums: a utopian and transcendental but also impossible figure, impossibly untouched since, in the discourse of realism, having a character is precisely to internalize, personify, or embody the social.

But it is, above all, the figure of the mother, or rather, the emphatic juxtaposition of the body of the mother and the social machine, that most powerfully condenses the relays we have been tracing: the relays between vision and embodiment and between social and natural ways of making persons. There is an extraordinary scene in the third chapter of Crane's story, when Maggie and her brother stare with fascination at "the prostrate, heaving body of the mother": "The urchin bended over his mother. He was fearful lest she should open her eyes, and the dread within him was so strong, that he could not forbear to stare, but hung as if fascinated over the woman's grim face. Suddenly her eyes opened. The urchin found himself looking straight into that expression, which, it would seem, had the power to change his blood to salt." And again, after the eyes of the mother have once again closed: "The eyes of both [Jimmie and Maggie] were drawn, by some force, to stare at the woman's face, for they thought she need only to awake and all fiends would come from below" (12–13). The body and "look" (in both senses) of the mother condense the charged relations of seeing and containment I have been indicating, relations, at once social and sexual ("all fiends would come *from below*"). Crane's stories return again and again to such encounters between the desire, or compulsion, to see and the fallen and heavily embodied body. What is compelling about such moments is, in part, the very compulsiveness of their repetition: the turns in positions of seeing and being seen that are also turns in positions of control. But what is perhaps most compelling about such moments is the thorough incorporation of relations of power into the movements of the eye and the dispositions of the body such that the problem of seeing and the problem of the body and the problem of slums each indicate the other, holding each other at such moments in fascinated suspension.

In short, the slums are personified in the figure of the mother, and that figure provides also a visual and corporeal model of the social. More exactly, for both Crane and Jacob Riis, for example, the slums are embodied in the body of a monstrously productive mother. The tenement in Crane's story, for instance, is "a dark region where, from a careening building, a dozen gruesome doorways gave up loads of babies to the street and gutter" (6).[11] The mother appears as a deeply embodied reproduction of the social, as social "forces" made visible and corporeal. And, like the fallen girl, the terrifying and fascinating "spectacle" of the mother is brought to book by the systematic machinery of justice and made legible as a police file or as a "case" study: "Peering down over his spectacles," the "police-justice" observes: "Mary, the records of this and other courts show that you are the mother of forty-two daughters who have been ruined. The case is unparalleled in the annals of this court, and this court thinks—" (41–42).

What the conversion of individuals into numbers and cases and the conversion of bodies into visual displays correlate are two of the crucial control-technologies of machine culture: statistics and surveillance.[12] Such a merger of optics and statistics posits a "remarkable parallelism between the desire to measure and the desire to look."[13] The merger between the desire to quantify and the desire to see makes possible, for example, the visual display of persons in the form of what the eugenicist Francis Galton called composite photographs or "pictorial statistics."[14] It makes possible as well the mergers of surveillance and statistics, the composite of visual displays and photographs, on the one side, and the graphs, numbers, fractions, and lines, on the other, that in combination make up Jacob Riis's account of how the other half looks. We might consider here the cases documented in the remarkable chapter of Riis's *How the Other Half Lives* that ends by picturing "the human files" and "human documents" of the prisons, hospitals, and workhouses of New York's prison island, counted bodies circulating through what Riis calls a social "machinery running smoothly."[15] The chapter begins by personifying the slums in these terms: "You and I are brothers. I am not more bankrupt in moral purpose than you. A common parent begot us. Twin breasts, the tenement and the saloon nourished us" (201). Riis's personification of the slums is thus also the social making or personification of persons. This is the understanding of persons as socially made that Riis here calls "growing a character," and that he elsewhere describes (in an autobiography titled, of course, *The Making of an American*) as the moment at which it becomes possible to speak of that deeply embodied other half achieving a "soul." The growing of persons is thus, for Riis, a function of the social calculus and a function of the seeing machine running smoothly.

The correlation of the visible and the calculable, and the understanding of persons as functions of the body-machine complex—as statistical persons—will be considered in some detail in the second half of this chapter. Here let me note that, in referring to a logistics of realism, I am not suggesting that such a logistics can itself be accounted for by referring these problems of embodiment and mechanism back to larger historical causes and forces. These relays and relations are in fact what they appear to be. They are not ultimately reducible to transparent symptoms or manifestations of deeper conflicts and contradictions but intrinsically and immanently intelligible. The realist account operates by way of the conflicts and reversals that constitute it. And a view of history (a Lukácsian way of seeing realism, for example) that accounts for persons, actions, or practices as instances of, or indeed as embodiments of, larger, determining historical causes or forces is itself a version of the conflicted discourse of embodiment that concerns me here (see also Part II).

Nor is this version of "the historical" the only way in which such a logistics continues to structure a wide range of contemporary practices and representations.[16] We might think here of the fascination with artificial reproduction; the ongoing renegotiations between constructivist and essentialist accounts of gender; the mutations of concerns about the subject and consciousness to concerns about embodied subjects and personification; or the fascination with the problem of making persons, biomechanics, and the body-machine complex in such hi-tech noir productions as James Cameron's and Gale Anne Hurd's *Terminator* or *Aliens*.

Or, to take up a final instance along these lines, it is in relation to the logistics of realism that we might consider Elaine Scarry's recent *The Body in Pain: The Making and Unmaking of the World*.[17] Scarry's extraordinarily compelling work concerns precisely the making-visible of interior states through a turning inside out of the body. The body figures alternatively, in her account, as "the physical basis of reality" and, in her euphoric invocation of a civilization without discontents, as something that can be made: "the remaking of the human body," as Scarry puts it, "is an ultimate aim of artifice" (253). For Scarry, the individual is a soul trapped in a body: "the body encloses and protects the individual within" (38). And the cultural aspiration toward disembodiment she describes issues in what she calls the "birth of the artifact" (259) from the "mother-lode" (137) of the body's interior. Moving back and forth between such an aspiration toward disembodiment and moments of incarnation, Scarry posits absolute oppositions between making and unmaking, pleasure and pain, civilization and power.[18] What these absolute oppositions and "monolithically consistent" (13) treatment ultimately render *in*visible are what might be described as the "modern" style of power: the relations of bodies and powers articulated, for instance, in a work Scarry's own argument in effect rewrites in reverse. I am referring to Foucault's analytic of the *productiveness* of modern relations of power and its implications. Specifically, Foucault investigates the ways in which techniques that make it possible to "see" induce effects of power, and the ways in which the discovery that the individual is something that can be made deploys "the natural machinery of bodies" most effectively not through torture, war, and the body in pain but through "a perpetual victory that avoids any physical confrontation."[19] Taking as its project a fabrication of the soul of the subject, this is a social machine at once corporeal and non-physical, "a power that seems all the less 'corporeal' in that it is the more subtly 'physical.' "[20] For Foucault, the body is not the prison of the soul, but the soul the prison of the body, and what Scarry describes as the "birth of the artifact" often resembles what Foucault calls the birth of the prison.

But it is just the possibility of such relations or resemblances that Scarry's absolute non-relation between making and unmaking disavows. Observing that terms such as "produce," "body," and "artifact" are common to models both of creation and of power, Scarry insists that "the overlapping vocabulary is itself the sign of how absolute the difference is between them, for they share the same pieces of language only because the one is a deconstruction of the other" (145). The overlapping of the terms of creation and the terms of power can only indicate an absolute difference because deconstruction, for Scarry, can have nothing to do with construction. The questioning of difference, the exigencies of reference, the relays that establish relations between apparently opposed registers: all are synonymous with destruction. Put simply, only by explicitly equating deconstruction with destruction can Scarry protect the monolithic differences upon which her single narrative depends. But despite reassertions of absolute difference that stand guard against the referential instability of shared pieces of language, it is in part the referential mobility of her own richly textured narrative, its absorption in relations of vision and embodiment, in the uncertain relations of artifactual and natural bodies and persons, that make for the aesthetic fascination of that narrative and also its implication in what I am describing as the cultural logistics of realism.

The Body-Machine Complex

I want at this point to take up the fascinations generated by the realist body-machine complex from a somewhat different perspective. I want again to take most of my examples from a writer who seems to be most explicitly and most obsessively engaged with these fascinations, Stephen Crane. But let me return to Crane through a brief consideration of several other texts that neatly broach, and extend, the cluster of concerns that the realist project insistently draws into relation.

George Eliot's *Daniel Deronda* (1876) and Edith Wharton's *House of Mirth* (1905) both centrally involve problems of personation and embodiment, and both take up these problems by way of a counterposing of embodied character and the automatisms that threaten to erode or to undo character and agency.[21] Both texts open by eliciting a series of questions about a central female character: Gwendolen Harleth in Eliot's novel, Lily Bart in Wharton's. What incites these questions, in both instances, is an indecision, a highly eroticized uncertainty, about the status of character and the embodiments and mechanisms that pressure that status. The opening scene of *Daniel Deronda* is of course a scene of gambling. And the opening questions about Gwendolen's form and expression, about her will or desire or force, are drawn into relation to matters of chance, probability, and perhaps, above all, a de-individualizing and inanimating *in*difference (one of the novel's complex words). The insistently *generic* description of persons in this opening scene—the "two serried crowds of human beings" who are "deeply engaged at the roulette-table" (35–36), engaged in what Eliot will later call "watching chances" (341); the description of characters "who had no individual physiognomy" (40); of "young men in general" (42); the repeated "measuring" of persons and chances—all are invoked, albeit not quite personified, in the "automatic voice" of the croupier, the "occasional monotone" of what Eliot calls "an ingeniously constructed automaton" (35).

There is more to be said about these terminal, terminator-like, descriptions of persons. But since I am most concerned with the network of relations established across these texts, let me turn to the somewhat similar scenario in Wharton's *The House of Mirth*. If the counterposing of watching numbers, figures, and chances and the watching of the figure of a young woman stimulates narrative interest in this opening scene, the questions that Lily Bart incites at the opening of her novel provide a certain commentary on the logic at work in such scenes. What attracts Selden to Lily in the initial scene is an uncertainty about whether to attribute Lily's acts to impulse or to intention, to chance or to calculation. "Her simplest acts," Wharton observes, "seemed to be the results of far-reaching intentions" (3). Lily possessed, for instance, "the art of blushing at the right time" (6), a spontaneity that yet seems calculable in terms of an "argument from design." It is the spontaneous or incalculable moment in these calculations—the power of converting impulses into intentions—of capitalizing on risk, that, above all, stimulates and attracts "interest" here.[22] At the opening of the second book of Wharton's novel, Lily Bart will be positioned watching chances at a gambling place. The juxtapositions of bodies and mechanisms of calculation or chance in these texts systematically evoke, alongside the automatistic character of the gambler, that other figure central to realist representation, the prostitute—as if the turning of, and the taking of, "tricks" were intimately related in the make-up of what Eliot calls "chancy personages."

Certainly, such a biomechanics of personation is not limited to the realist or naturalist novel. We might consider, for instance, the work of late nineteenth-century sociologists and psychologists. I have in mind particularly the work of William James, whose *Principles of Psychology* (1890) treats the "adjustment of inner and outer relations" and the character of "machine-like yet purposive acts" in terms of an interplay between the spontaneous and the incalculable, on the one side, and the fatal and machine-like, on the other. James's psychology, in its accounts of "automaton-theory," and of the mind as a "physical machine" with interests, represents the interests that, in effect, personify the physical machine in these terms: "The performances of a high [that is, human] brain are like dice thrown forever on a table. Unless they be loaded, what chance is there that the highest number will turn up oftener than the lowest? All this is said of the brain as a physical machine pure and simple. Can consciousness increase its efficiency by loading its dice? Such is the problem."[23]

We might consider as well the work of James's near-contemporary, Freud, and not merely because the interests generated by the body-machine complex are, as we have seen, insistently eroticized and specifically gendered, and not merely because these texts, Eliot's, Wharton's, and Crane's, lend themselves so readily to psychoanalytic attention. Crucial here are the ways in which psychoanalysis's separations of the psychic from the physical, of the inside from the outside, operate through a competition between determinism and chance. That is, we might consider the ways in which psychoanalysis, in its fascination with the competition between determinism and chance, and between the psychic mechanism and bodily insistences, shares in, even incites, the erotics it claims as its "object," shares in and forwards precisely the fascinations of the realist complex I have been sketching.[24]

But for the moment I am more interested in a related but somewhat different interface between realism and psychoanalysis, a final relay in this configuration of relations. Psychoanalysis, as Jacques Derrida has observed, inclines toward "increasing reserves of random indetermination" even as this apparent randomness is placed in the service of a compulsive determinist interpretation. To paraphrase Freud, "(real) chance" exists in "external events" but not "in internal (psychical)" events. There is thus no chance in the unconscious.[25]

Derrida's inside critique of psychoanalytic determinism, it will be recalled, centers on Lacan's insistence that a letter always arrives at its destination, that there are, as it were, no dead letters in psychoanalysis. But it is precisely a discovery about dead letters that provided the take-off point for one of the most dominant discourses of the nineteenth century, a discourse about chance, agency, and causation that underwrites nineteenth-century biopolitics and styles of realism both: the first well-documented instance of a statistical order that could not plausibly be interpreted in terms of conscious or purposive intentions was Laplace's announcement in the *Philosophical Essay on Probabilities* that the number of dead letters in the Paris postal system remained constant from year to year.[26] The discovery of such regularities irreducible to individual intentions, and the generalization of the statistical method, are closely bound up with the emergence of a biopolitics of populations, with the possibility of a social science. At the same time, the de-individualizing tendencies of statistics provide models of individualization: models for the generic, typical, or average man, for what I have been describing as the production of individuals as statistical persons.

"There were men who nearly created a battle in the madness of their desire to see the thing. . . . Meanwhile others with magnificent passions for abstract statistical information were questioning the boy. 'What's his name? Where does he live?' " (Crane, 602). The thing that the faceless crowd of individuals in Crane's story "When a Man Falls, a Crowd Gathers" desires to see is the "inert" and "rigid" but living body and "livid visage" of a man who has collapsed onto the sidewalk, apparently in a kind of seizure. A little later on, Crane writes, "the peering ones saw the man on the side-walk begin to breathe, with the strain of overtaxed machinery" (603). The impassioned struggle to see, to see inanimated bodies or bodies whose animation has been suspended, is, as already noted, an obsessively recurrent scenario in Crane's work. So too is this "magnificent passion" for statistical information: the powerful investment of affect in the abstract, generic, or statistical and the rendering of the most highly individualized and particularized (What's his name? Where does he live?) as, simultaneously, abstract and general. One finds in Crane, and in late nineteenth-century discourse generally, something like a passion for *cases*, in several senses, and for case histories.[27] Finally, there is the close and violent relation in these accounts of personation—one that surfaces only momentarily in this story—between the docile body and a certain mechanics of biomechanics ("the strain of overtaxed machinery").

Crane's stories are populated if not quite peopled by statistical persons: by the figure of the gambler and the prostitute and also by that emphatically embodied and paradoxically *particularized type* that is, at once, the intended product and the apparently accidental by-product of war—the wounded soldier or *casualty*. Crane's statistical persons, often unnamed (I am tempted to say "only nominally named"), are at once radically embodied and strangely generic. They are most often positioned at a crossing point between the accidental or casual and the causal. The stories obsessively return to numbers, counting, and calculation ("the peering *ones*," for instance). They obsessively return, above all, to an interface of vision and embodiment figured as a violent exchange (an *excruciating* exchange—literally, a crossing point) between the body and the machine.

The relations between the realist investment in seeing and its biomechanics can be clarified through a series, or in fact two series, of examples. Crane's thematics are almost schematically visible, almost parodically foregrounded, in his very late story "Twelve O'Clock." Entering a hotel, "a notable place" owned by a man named Placer, a cowboy is astonished—rendered "stupefied, glassy-eyed"—by "the machinery of a cuckoo-clock": by, as he describes it, "a clock with a wooden bird in it," a "bird cage" that tells time (831–32). A sequence of charged encounters, scarcely motivated in any manifest way, follows, until Placer, who is throughout represented as "writing in his ledger," takes up a revolver in each of his "incompetent hands" and is shot behind the "absurd pink counter" at which he has been writing: "Placer fell behind the counter, and down upon him came his ledger and his inkstand, so that one could not have told blood from ink" (835).

Recalling the terms of Maggie's wooden doll that talks, the story centers on a series of violent intersections between animism and mechanism or automatism. These are figured most prominently in terms of the animate machine of the clock and a deathly mingling of writing matter and bodily matter, ink and blood. In a recent and richly provocative essay on Crane to which my account is already indebted, Michael Fried argues that Crane's stories recurrently involve a thematization of writing, or, more explicitly, of the scene of writing.[28] Fried emphasizes and inventories the ways in which Crane's repeated scenarios, particularly the upturned face and fallen body, might be read as invocations of the scene of writing. Crane's writing, Fried also suggests, enacts a conflicted and obsessive encounter between the at once physical and immaterial act of putting letters and words on paper and the powerfully animating or animistic effects of this act. In short, the scene of writing is at once invoked and necessarily repressed in Crane's work; and it is necessarily repressed, or violently displaced, because narration, personality, and characterization are everywhere threatened by their exposure as merely effects of certain practices of writing.

Fried's account of "the scene of writing," in these terms, is not entirely unfamiliar. Despite differences in emphasis, and particularly despite the strikingly decontextualized and autoreferential quality of Fried's account of Crane, one is reminded here especially of the "deconstructive" accounts of writing and disfiguration provided by Jacques Derrida and Paul de Man; at the same time, this way of proceeding makes it possible for Fried to specify, with a brilliant intensity, what John Berryman identified as the "lunar" effect and strangely autogenic character of Crane's writing.[29] What Fried leaves unexplained, however, in pointing to the conflicted or threatening character of writing in Crane's work, is the specific violence generated in such realist texts, and in Crane's writing in particular, by the obsessive turn to the scene of writing. And, along what I want to suggest are the same lines, his account does not seem to register (or registers only by displacing) the specificity and "historical specificity" (this phrase will need unpacking) of the body-machine complex everywhere visible in Crane's stories. Put simply, Fried's decontextualizing account does not quite explain, or renders too quickly transparent in its explanation, what might be called the surfacing of the body and the machine in Crane's writing. I am suggesting that, despite (or as a consequence of) the extraordinary attention, in that account, to what Crane's writing literally "looks like," these graphic cases too quickly become a transparency through which a general thematics of writing is read. That is, the specific subject or surface matter—torn and automatistic bodies, mechanisms and pieces of technology, natural and mechanical inscriptions across bodies and landscapes—are too quickly reduced to a writerly self-absorption in the scene of writing. Hence that account does not quite register what I have described (in the introduction to this study) as the *abstract materialism* of realist writing and its effects: the specific ways in which the abstract materiality of turn-of-the-century realist writing and turn-of-the-century realist writing-technologies communicate with each other.

The scene of writing is arguably an inevitable thematization and, as it were, ready-to-hand version of the logic of embodiment for the writer (as in the slight story "Twelve O'Clock" from the "incompetent hand" of the dying Crane). It is also a perhaps inevitable version (in the interface of head, hand, and the automatisms of language) of the general conflicts between ideality and materiality, and of action and automatism, traced here. But if the scene of writing is not just one version of these conflicts among others, neither is the scene of writing the ground of these versions, a "source" that proliferates its images throughout Crane's stories.

Put somewhat differently, it is not exactly images of writing or the scene of writing or a thematization of writing but, as Fried most incisively characterizes it, the becoming-visible of writing as writing that must be considered here. But the becoming-visible of writing as writing—the realist/naturalist *writing of writing*—indicates something more here than the writer's self-absorption.[30] What must be considered is a programmatic equation or identification: the perfect "fit" between technologies of writing and the body-machine complex and the perfect "fit" between the ontology of writing-in-general and an historically specific biomechanics. What this amounts to, in the most general terms, would be the perfect coincidence of the apparently incompatible claims of a radical formalism and a radical historicism. What from one point of view, in the realist/naturalist writing of writing, looks like an ontology of writing looks, from another, like a sociology of biomechanics, statistics, regulative social surveillance, styles of automation: an *ontology* of writing that is indistinguishable from realist *sociology*. It is this identification of writing and social physics as two versions of the same thing (each as *une langue inconnue* of the other) that in effect terminates the realist project: or, better, produces the terminator-version of the realist project sometimes called naturalism.

Another way of saying this is that the perfection of realism would be a perfect referentiality: *bodies and matter writing themselves*. But if the perfection of realism would be its perfect referentiality, the identification of technologies of writing and "the subject" of writing as two versions of the same thing would render the realist project purely tautological. The perfect referentiality of realist writing would thus be its perfect self-referentiality: the realist project of accounting for persons as effects or products of the machine, as typical and predictable functions of a social biomechanics, would be identical to accounting for persons and personation as effects of writing. These are the anxieties and also the appeals registered, for example, in the relays between calculation and representation and in the representation and calculation of individuals as statistical persons.

It should by now be clear that in referring throughout to "the realist project," I am referring to a *model* of realism toward which particular realist practices gravitate; and, in fact, the power of the model at work in realism, which I will take up briefly at the close, is crucial here. But first I want to set out more specifically several more cases in Crane's practice, and to set them out with minimal comment, in part because their import should by now be in evidence and in part because it is what connects these cases that most concerns me here. Crane's emphases are nowhere more evident than in the stories that pivot on watching chances or gambling: for instance, "The Five White Mice" and "The Blue Hotel."

The central moment of "The Five White Mice" is a moment of suspended action. The "straight line" towards this moment involves a series of accidents set in motion by a throw of dice (the five white mice are part of a gambler's rhyme, "the five white mice of chance"). An unintended "jostling" of a man in one group by a drunken man in another motivates one "who up to this time had been an automaton" and induces "a sort of mechanical fury" in another (767, 766). (It is perhaps worth recalling here that "automaton" can mean either "a thing regarded as capable of spontaneous motion or action" or what is just the opposite, "a person who behaves mechanically . . . following a routine without active intelligence.") The subsequent face-off is suspended first by a fantasy about writing and second by a confrontation between body and machine, both of which condense the effects of animism and inanimation or mechanism evoked by these confrontations in Crane. The gambler fantasizes his own death by imagining its report in a telegram "written in a careful hand on a bit of cheap paper. . . . But they are often as stones flung into mirrors, these bits of paper upon which are laconically written all the most terrible chronicles of the times" (768). In *Maggie*, an object is thrown into a pyramid of shimmering glasses and "Mirrors splintered into nothing" (37). In "Five White Mice," this piece of writing—an obituary that is also a sort of suicide note—appears as a mirror (or mimesis or personation) that splinters into nothing. After fantasizing this paper written in a careful hand, the gambler takes up in a "series of movements" and as if "unconsciously" a revolver which he "feared that in his hands would be as unwieldy as a sewing-machine. . . . But at the supreme moment the revolver came forth as if it were greased and it arose like a feather. This somnolent machine, after months of repose, was finally looking at the breasts of men" (769). The (non)story ends with this short paragraph: "Nothing had happened" (771). The confrontation between the body and the machine, the violent, even suicidal, conflict between writing as writing and as the representation of an action: these are the relays that structure and suspend (the possibility of an) action in the story. The rapid alternations between deep embodiment and disembodiment (between weightiness and weightlessness), between animate machines and persons whose animation is suspended, between bodies made visible and vulnerable to "looking" machines: these are the tensions that Crane's writing of writing sets in motion.

"The Blue Hotel" also narrates a chain of incidents that follows from a scene of gambling. A character called "the Swede" shows "signs of an inexplicable excitement" (801) and seems convinced that he is about to be killed. In fact, he manages to bring about his own death through what can only be called a suicide by proxy—suicide figuring, of course, as the preferred method of "becoming a statistic" in the nineteenth century, as the suicide note only becomes a popular genre in the last century. The Swede instigates a fight over a game of cards and subsequently provokes his murder at the hands of a professional gambler: "and then was seen a long blade in the hand of the gambler. It shot forward, and a human body, this citadel of virtue, wisdom, power, was pierced as easily as if it had been a melon" (826). The paradoxical notion of the "professional gambler" is perhaps the perfect figure of the combination of chance and determination or expertness, hence the perfect emblem of a culture of professionalism. In a brief scene that functions as an epilogue to the story, a character called simply "The Easterner"—a likely nickname for Crane himself—"enter[s] with the letters and the papers," and comments on the professional gambler's paradoxically passive action in these astonishing terms: "We're all in it! This poor gambler isn't even a noun. He is kind of an adverb" (827).

It is not merely that this description converts the apparent agent of action into a mere modifier of an action. Nor is it only that "the adverbial form," as Donald Davidson has suggested, "must be in some way deceptive." The adverbial form must be in some way deceptive in that the form may imply that intentionality or agency is something extra that may be added to or subtracted from an action, and hence that actions are potentially separable from agency or intention (that is, not actions—the acts of an actor—at all). As Davidson puts it, "doing something intentionally is not a manner of doing it. To say someone did something intentionally is to describe the action in a way that bears a special relation to the beliefs and attitudes of the agent. . . . But of course to describe the action of the agent as having been caused in a certain way does not mean that the agent is described as performing any further action," that is, the "further" action of intending to act.[31]

Crane's tendency to substitute adverbial forms for direct verbs of action has frequently been commented on. More generally, the well-marked tensions between agency and chance or determinism, the becoming problematic of what George Eliot in *Daniel Deronda* calls "the logic of human action,"[32] have frequently been noted in realist and naturalist writing. My point is that, in Crane's story, the terminal logic of human action entails the reduction of the "human body" to mere matter and, collaterally, the reduction of human action to the names of parts of speech. The becoming visible of writing as writing, "in this case" of letters and papers, produces "inexplicable" effects of violence and impersonation.

Crane's writing thus typifies typicality and the becoming abstract and general of particular human actions, not least the act of writing itself. Representing particular actions as cases and particular individuals as chancy and statistical persons, Crane's practice of writing seems to represent, or to make itself visible, as at once a terminal instance of a conflicted "writing in general" and of the historically specific network of discourses and practices in which such conflicts become pressing or appealing. The possibility of representation and personation here cannot be separated from these specific tensions between materiality and typicality and between the particular and the general. It cannot be separated from them because, as we have seen, representation and personation in realist writing (and also in the contemporary continuations of the realist project often described as post-realist) everywhere *realize* these tensions.

Crane's *The Red Badge of Courage* (1895) foregrounds this cultural logistics at every point: in the indecisive relations of chance and action—the "mental calculations" of the casualty; in the generation of effects of animism and automatism—through collisions of bodies and machines; and in the catalyzing of these relations—through a desire to make visible that triggers the making-visible of writing itself. I want to postpone a more detailed look at *Red Badge* until the final part of this study, and my account there will take up this logistics from a somewhat different perspective. This in part because moments such as these scarcely require much interpretive pressure: "Once the line encountered the body of a dead soldier. He lay on his back staring at the sky. . . . The youth could see that the soles of his shoes had been worn to the thinness of writing paper, and from a great rent in one the dead foot projected piteously" (101–2). The projection of the body through a "sole" worn to the thinness of writing paper, the very surfacing of the pun and the effect of the pun as a materialization both of writing and of personhood, are evident enough. It is, again, the correlation of the writing of writing and the body-machine complex—a correlation nowhere more powerfully instanced than in Crane's story of the disciplinary machinery of the *corps*—that I have been tracing here.

Briefly, effects of psychology and personation are produced in response to these sights in the novel, to sights of lines encountering bodies. And *Red Badge* makes clearer the eroticized and gendered character of these effects. To anticipate a bit: *Red Badge* in effect tells two stories at once, a love story and a war story. On the one side, there is an "inside" story of male hysteria, registering fears of unmanning and infantilization in the face of threats of bodily dismemberment that are frequently localized (if that is the right word) in well-marked fantasies of castration and maternalized engulfment; on the other, an "outside" story of social discipline, mechanization, body counts, and the industrial and *corporate* disarticulation of natural bodies (the production of what Crane tells "the body of the corps"). The exact interchangeability of these two stories (the psychoanalytic and sociohistorical) points again to a terminal point in the realist project: to the surfacing, in this highly charged renegotiation of inside and outside, of what I have described as the realist tautology.

Crane's story everywhere evokes the idiom of the probable casualty and other chancy persons, and not merely in the repetition of the terms of gambling ("you bet") and the calculus of probabilities ("what d'yeh think th' chances are?" [161]). Charles Sanders Peirce, the pragmatist philosopher and longterm employee of the U.S. Government in the Coast Survey (his job was measuring and research in measurement devices) observed, in "The Doctrine of Chances" (1878), what he called the "logicality" of "the heroism of self-sacrifice": "All human affairs rest upon probabilities . . . Sometimes we can personally attain to heroism. The soldier who runs to scale a wall knows that he will probably be shot, but that is not all he cares for. He also knows that if all the regiment, with whom in feeling he identifies himself, rush forward at once, the fort will be taken. In other cases we can only imitate the virtue . . . our logician would imitate the effect of that man's courage in order to share his logicality."[33] This is, in brief, the courage of statistical persons, acting to share in the logicality of cases, shares, and chances.

"The battle was like the grinding of an immense and terrible machine to [Henry]. Its complexities and powers, its grim processes, fascinated him. He must go close and see it produce corpses. . . . [S]uccess for the mighty blue machine was certain . . . it would make victories [produce corpses] as a contrivance turns out buttons" (129, 146). The mechanical reproduction of persons appears here only in the inanimated, or for Crane, semi-animate body of the corpse, the person who has become, simply, a body.

Earlier, Henry, panicked by the fear of and also desire for bodily injury, is represented as "picking nervously at one of his buttons. He bent his head and fastened his eyes studiously upon the button as if it were a little problem" (132). The blue machine that makes buttons and this little problem of the button that Henry fingers may be linked by the joke current in the nineteenth century and beyond (it turns up, for instance, in Pynchon's *V*) and which appears most prominently in a novel that anticipates, in different form, many of the concerns about personation, embodiment, and dismemberment displayed in Crane's work.[34] The joke is in Melville's *Moby Dick*: "But, unscrew your navel, and what's the consequence?" The consequence, of course, is that your bottom falls off. And the point of the joke about the button and the body's falling apart neatly conjoins the rival accounts of making persons—the mother and the machine—noted at the start. This is perhaps the little problem of the button on Henry's belly.

Working Models

I want to close by very briefly taking up what I earlier called the power of the model in realism. I have to this point indicated some of the conflicted forms the logistics of realism takes. It may now be possible to suggest some of the ways in which these forms, and the fascinations they generate, have been channeled into certain practices of power and control. But I don't mean here to imply an opposition or separation of abstract "forms" and material "practices." Rather, it's the fascinations generated by the relays between the formal and the material, the abstract and the practical, the generic models and the concrete individuals that, we have seen, make up statistical persons—and, I want to suggest, that make up and deploy the paradoxical *working models* of machine culture.

Jacob Riis ends *How the Other Half Lives* not with the counterposing of the mother and the machine discussed earlier, but, instead, with a series of graphs, maps, line drawings, statistics, and models for model tenements. This is a text preoccupied with drawing lines, and not least color lines (race or "color" providing an exemplary instance of the realist desire to make interior states and social forces visible and mutually intelligible). Riis's photographs routinely juxtapose rounded or shapeless human forms against strongly lined, almost gridded, backgrounds. The major target of Riis's exposure of the other half was a section of the slums called "the Bend," and one of the first effects of his documentation was a removal of the Bend and renewal of the area along the lines, straight lines, Riis proposed. In all of these instances, one finds a juxtaposition, or system of flotation, by which a normative model is "floated" in relation to a material practice or body. The force of the model as *working model*—formal and material at once—effectively incites the regulation, containment, or mobilization Riis calls "reform."

Here are three examples from Crane's writings. The first, from the extraordinary story "Death and the Child," juxtaposes drawings to wounded bodies, producing an eerie, sanitized typicality. The narrator is describing a line of wounded soldiers: they "were bandaged with the triangular kerchief upon which one could still see through bloodstains the little explanatory pictures illustrating the ways to bind various wounds. 'Fig. 1.'—'Fig. 2.'—'Fig. 7.' " (947). The superimposition of the visible and the calculable, representation and quantification, physical bodies and abstract models: these are the working models, living diagrams, and unnatural nature that, as we have seen, correlate wounded bodies and writing bodies in machine culture.

The second example, from Crane's *Maggie*, describes a scene of consumption in these terms: on a sideboard behind a shining bar

> rested pyramids of shimmering glasses that were never disturbed. Mirrors set in the face of the sideboard multiplied them. Lemons, oranges and paper napkins, arranged with mathematical precision, sat among the glasses. Many-hued decanters of liquor perched at regular intervals on the lower shelves. A nickel-plated cash register occupied a position in the exact center of the general effect. The elementary senses of it all seemed to be opulence and geometrical accuracy.
>
> Across from the bar a smaller counter held a collection of plates upon which swarmed frayed fragments of crackers, slices of boiled ham, dishevelled bits of cheese. . . . An odor of grasping, begrimed hands and munching mouths pervaded. (33–34)

The flotation of a geometrical model, at once formal and material, across from practices of consumption has the effect of inciting and directing that consumption. It functions as a working model, in the manner of the modern forms of advertising that emerged in the 1890s: for instance, the geometrical typicality of the model or manikin, the abstract body of the consumerist doll or automaton. Typically, Crane signals thick materiality by materializing writing (*fr*ayed *fr*agments . . . dis*h*evelled bits of *ch*eese . . . an o*d*or of *gr*asping, be*gr*ime*d* han*d*s and *m*unc*h*ing *m*ouths perva*ded*); typifying typicality, he counterposes effects of geometry and deeply embodied bodies, here organized and registered by the machinery of consumption, occupying "the exact center." Effects of personation ("Mirrors set *in the face* of the sideboard," for instance) are counterposed to the "general effect" of regular series and intervals. Hence "swarming" body parts are arranged into the consumer body. This is what consumption here looks like: the pleasures of systematically coordinated and managed bodies (see Part V) set in motion by the "tinkling of the apparatus" (see Introduction)—the mechanism that incites and measures, registers and personates "the consumer."

My final example is Crane's description of the engineer in his article "In the Depths of a Coal Mine" (see Figure 1). The article appeared in *McClure's Magazine* in August, 1894. In the same volume there is an article (by a professor of physiology at Harvard) titled "Are Composite Photographs Typical Pictures?" (see Figure 2).[35] The piece concerns the method invented by the eugenicist and human engineer Francis Galton of "employing the photographic camera to combine the features of a number of individuals upon the same sensitive plate, thus producing a typical portrait of the group by [as Galton puts it] 'bringing into evidence' " all common traits "and leaving 'but a ghost of a trace of individual peculiarities.' " The composite photograph—Galton called them, as we have noted, "pictorial statistics"—provided the necessary visual analogue of a social typology and of deviations from the type, merging looking and measuring in a standard, and standardizing, schema. Combining the visual/bodily and the ideal, the composite photograph epitomizes the interface of models and practices I have been indicating. The composite picture, like the "composite monster" of the "war machine" (110) in Crane's *Red Badge*, links natural bodies and the machine process. The "brilliant measurings of his mind" (195) in Crane's novel, the "types of faces" (191), the repeated "deep calculations of time and distance" (180), the "figurings" and "diagrams" and "dissections" (205): in all these cases, the bearing of standards and uniforms across a violent landscape makes visible standardized and uniformed bodies that are also torn and violated bodies (see Part V).

The human engineer appears (as Santayana characterized the typical American at the turn of the century) as "an idealist working on matter"; the engineer, that is, mediates between models and practices,

Figure 1. The Engineer, from "In the Depths of a Coal Mine," illustrated by Corwin Linson.

Figure 2. The Composite Photograph, or, Pictorial Statistics.

embodies and personifies "transcendental realism."[36] Crane's picture of the engineer is also a picture of the new science of human engineering:

> We were on our swift way to the surface. Far above us, in the engine-room, the engineer sat with his hand on a lever and his eye on the little model of the shaft wherein a miniature elevator was making the ascent even as our elevator was making it. In fact, the same mighty engines give power to both, and their positions are relatively the same always. . . . My mind was occupied with a mental picture of this faraway engineer, who sat in his high chair by his levers, a statue of responsibility and fidelity, cool-brained, clear-eyed, steady of hand. His arms guided the flight of this platform in its mad and unseen ascent. It was always out of his sight, yet the huge thing obeyed him as a horse its master. When one gets upon the elevator down one of those tremendous holes, one thinks naturally of the engineer. (614)

Crane's fascinations are nowhere more provocative or more graphic than in this astonishing "mental picture" of the power of the little working model, and of the virtually automatistic engineer encased in the machine he is controlling. This little model and immobilized mover, provide—in the technology coordinating head and hand, body and machine, mastery and the "mad and unseen"—the very model of the body-machine complex and its logistics.

Part IV

The Still Life

The Aesthetics of Consumption

One of the most evident paradoxes of the insistently paradoxical notion of a "culture of consumption" is the manner in which a style of life characterized by its excessiveness or gratuitousness—by its exceeding or disavowing material and natural and bodily needs—is yet understood on the model of the natural body and its needs, that is, on the model of hunger and eating. Hence if, as Jean Baudrillard argues, "it is necessary to overcome the ideological understanding of consumption as a process of craving and pleasure, as an extended metaphor of the digestive functions," it is nevertheless the case that such an understanding continues to govern accounts of consumption, both "for" and "against."[1] There is certainly nothing unusual about such a linking of political economy and physiology. The notion that the body and the economy indicate each other is a commonplace of economic thinking from Aristotle to Malthus or Marx to the present.[2] But it is precisely the anti-natural and anti-biological bias of the culture of consumption—we might say the *sheer culturalism* that marks the discourse of consumption—that makes such a propping of consumption on the body seem paradoxical or contradictory.

Yet this paradoxical relation to the body perhaps represents something more than an "ideological understanding" or misunderstanding. For Baudrillard, for instance, it is as if the extension of the pleasures of the body to its representations, the extension of biology into ideology, is itself "a process of craving and pleasure," a sort of addiction to the pleasure of overcoming the body that, in turn, "it is necessary to overcome." The "culture of consumption" would, on this view, seem to have something like the same relation to culture "as such" as the forms of representation called pornography have to sex "as such." In the turning away from the body to representation, the story goes, in the turning away or perversion from need to want, in short, in the turning away from nature and necessity to pseudo-needs and unnatural or artifactual wants, both the culture of consumption and pornography—and they are, for this reason, if not exactly inseparable, exactly congruent—represent a fall into representationalism that is marked, at least retroactively, not by a desire for "the thing itself" but for its representations or substitutes. But if, on this account, consumption is the pornography of culture as pornography is the form of erotic life proper to consumer or commodity culture, putting matters this way already indicates something of the difficulty with such an account. It indicates something of a difficulty because the turning away from nature and necessity, the propping of representations on the body, the "necessary perversion of desire away from biology/need" to the artifactual and representational—all might be taken to define both culture *as such* and sexuality *as such*.[3] From this point of view, the critique of the culture of consumption in terms of its gratuitousness and unnaturalness amounts to a critique of culture in general: not a critique of the culture of consumption but a critique of culture *as* consumption.

My interest here is not in taking one side or the other of the general critique I have briefly set out. In fact, such a critique seems to me too general and abstract either to agree with or to disagree with. I am interested, that is, less in just saying no to consumption or just saying yes to it, than in considering some of the anxieties and appeals that the overly abstract form of the consumption debate at once intimates and obviates. I have earlier traced some of the relations between the bodily and the economic in market culture (the relations of "physical capital") and the making of individuals in machine culture (the making of "statistical persons"). I want here to take those accounts a step further, particularly by way of a reconsideration of the problems of the body and the melodramas of uncertain agency precipitated by what we might call an *aesthetics of consumption*.

Two cases may serve to epitomize, from opposite sides and in very preliminary form, what these problems of agency and embodiment look like. Thorstein Veblen, in an article called "The Economic Theory of Woman's Dress," which appeared in *The Popular Science Monthly*, December 1894, rehearses several of the topics that will center, five years later, what is still in many ways the governing account of the rituals of consumer culture, *The Theory of the Leisure Class* (1899).[4] Veblen's theory of woman's dress anticipates, for instance, the notion of the female body as a sort of leading economic indicator of consumer culture: "the function of woman, in peculiar degree, to exhibit the pecuniary strength of her social unit by means of a conspicuously unproductive consumption of valuable goods" (200). It anticipates also the manner in which the unmaking and remaking of the natural body carries the weight of consumer society's bioeconomics. As Veblen puts it, "voluntarily accepted physical incapacity argues the possession of wealth" (203). But such an attenuation of the natural body is linked to the aesthetic remaking of bodies and persons. The "mixture of the aesthetic and economic" in fashion draws into relation the consumerist canons of "good form" and "visible expense." Beyond that, the aestheticization of the body makes possible what Veblen invidiously describes as a pro-choice or voluntaristic principle of identity or personhood. Veblen concludes his article in these terms:

> The theory of which an outline has now been given is claimed to apply in full force only to modern woman's dress. It is obvious that if the principles arrived at are to be applied as all-deciding criteria, "woman's dress" will include the apparel of a large class of persons who, in the crude biological sense, are men. This feature does not act to invalidate the theory. A classification for the purpose of economic theory must be made on economic grounds alone, and can not permit considerations whose validity does not extend beyond the narrower domain of the natural sciences to mar its symmetry so far as to exclude this genial volunteer contingent from the ranks of womankind. (205)

The replacement of "the crude biological sense" by "economic grounds" thus appears as the replacement of the natural body by the symmetries or logic of economic theory and, correlatively, of natural identities by "volunteer" ones.

It is not merely that the fashion industry provides the most evident instance of what Elizabeth Wilson calls "the long running dialogue between the artificial and the natural in western industrial society."[5] Nor that the fashion industry most conspicuously incites and deploys the tensions between generic and standardized models, sizes, and styles, on the one side, and personalizing deviations and desires, on the other, that make it possible for fashion to be at once an object of desire and an industry. As Georg Simmel puts it, "the vital conditions of fashion"—by which he means the vital conditions of the modern "style of life"—represent "the adaptation of the special to the general," an adaptation that coordinates the "two antagonistic principles" of uniformity and differentiation.[6] The adaptation of the special to the general is, in short, what allows statistical persons to inhabit what Daniel Boorstin identified as "consumption communities" and for the consumption of differences nevertheless to allow for the democratized experience in what Boorstin also identified as "statistical communities."[7] This is, in brief, the rhythm implicit in the possibility of "sharing the experience" itself.

The making of statistical persons thus renovates what might be called the privilege of relative disembodiment in market culture. Such a privilege of relative disembodiment or relative weightlessness is one sign of the *aestheticization* of the natural body in market culture, and such an aestheticization of the body one sign of the achievement of personation through practices of consumption.[8] These relays between aesthetics, consumption, and personation are richly exemplified in an extraordinary sequence early on in Rebecca Harding Davis's novel of industrial life, *Margret Howth: A Story of To-Day* (1862).[9] The passage depicts the "queer little body" (57) of a female huckster or, rather, juxtaposes her "distorted" (65) body to the wagon-load of goods "in the middle" of which she sits. What the mingling of body and market goods foregrounds is the transcendence of overly deep embodiment through a rudimentary aesthetics of color, arrangement, and composition:

> Lois shook down the patched skirt of her flannel frock straight, and settled the heaps of corn and tomatoes about her. . . . The flannel skirt she arranged so complacently had been washed until the colours had run madly into each other in sheer desperation . . . the masses of vegetables, green and crimson and scarlet, were heaped with a certain reference to the glow of colour, Margret noticed, wondering if it were accidental. (64)

The passage traces stages in a process of composition: the process extends from the distorted little body of the huckster, through the mad wash of colors and the objects of consumption about her, and, by way of the arrangement of mass and masses into form, to the composition of something like a still life.

The composition of the still life links aesthetics and the market. But the artistic "glow of colour" substitutes also for the overly deep embodiment represented by another sort of color: "Nothing but the livid thickness of her skin betrayed the fact that set Lois apart from even the poorest poor,—the taint in her veins of black blood" (56). Imagining black skin as thick skin imagines race in terms of the too thickly embodied body. It imagines also a proximate relation between the racialized body and the industrialized body—Lois's tainted body is seen also as a product of the mills, where she "just grew": " 'It was th' mill,' she said at last. 'I kind o' grew into that place in them years: seemed to me like as I was part o' th' engines, somehow. Th' air used to be thick in my mouth, black wi' smoke 'n' wool 'n' smells . . . th' black wheels 'n' rollers was alive" (69). Consumption and racialization thus appear as the two sides of a single formation, "opposed," that is, as the two sides of a horseshoe. The aesthetics of consumption appears as the antidote to overly deep racial and class embodiments, conferring a privilege of relative disembodiment. That privilege appears here as the conversion of persons and things into artifacts and, more exactly, into the commodity form of the still life.

In what follows I will be concerned with the radical entanglement of consumption and racialization (what I will take up as the *miscegenation of nature and culture*) and, collaterally, with the aestheticization of the natural body that market culture promises (what I will take up centrally in relation to the visual representation of objects of consumption—particularly, the still life). My examples are drawn largely from Rebecca Harding Davis's remarkable story of the body-machine complex, *Life in the Iron Mills; Or, The Korl-Woman,* and, in part, from Veblen's accounts of the melodrama of uncertain agency in machine culture. But before returning to Davis and Veblen, it may be useful to set out in a bit more detail some of the ways in which recent cultural criticism continues to inhabit, or reoccupy, the premises of an aesthetics of consumption.

Agoraphilia

If fashion "links the biological body to social being," it also represents, in the conversion of bodies into styles (the understanding of the body as something that can be made), a strictly culturalist transcendence of the body and its limitations.[10] One might instance here the only superficially contradictory convergence of American "physical culture" and capitalist asceticism (no pain, no gain) or the anorexic "ideal of being thin, of being without a body" (never too rich or too thin).[11] These are the intrinsically paradoxical expressions of an asceticism that denies what counts as natural need for the pleasure of a pain that can't be reduced to the natural or the necessary: the disinterestedness of an aesthetic purified of the *interestedness* of pleasure itself.

We might consider here also the generally lurid fascination with which the forms of popular culture register these culturalist preferences for the artificial and the artifactual. These anti-natural and anti-biological biases are perhaps clearest in the fascination with the becoming-unnatural of biological reproduction, in, for example, the accounts of reproduction and reincarnation that we have seen govern the perverse obstetrics of the naturalist machine. "Science Nears the Secret of Life . . . A Long Step Towards Realizing the Dream of Biologists, 'to Create Life in a Test Tube.' " These are the headlines from an issue of the *Chicago Sunday Tribune* dated November 19, 1899. (The subject is Jacques Loeb, whose application of the principles of scientific engineering to the life process, at the turn of the century, paralleled Frederick Winslow Taylor's application of the principles of scientific management to the work process.[12]) It's not hard to see how contemporary conflicts about creationism, sociobiology, and reproductive technologies reoccupy the terms of later nineteenth-century conflicts about evolution and devolution, eugenics and human engineering; such conflicts continue to govern the form reproductive rights debates have taken, most visibly in the problem of redrawing the vague and shifting line between "the natural" and "the cultural" and in defining what counts as a person and apportioning rights to embodied persons (that is, the "choice" between one person with two bodies, or two persons in one body).

What one discovers, across these conflicts about persons, bodies, and technologies, is both the continuing fascination of a certain miscegenation of nature and culture and an insistence on the segregation of these sieve-like terms. I will be returning to the fascination generated by the miscegenation of the natural and the cultural and the panic that takes the form of a violent movement of segregation: the politics of culturalism articulated along these uncertain and therefore well-policed lines. Clearly, it's not a matter of transcending these contradictions and not least because this would be to ignore or to deny their *appeal.* Recent accounts of the culture of consumption, and recent "culturalism" or new historicism generally, have tended to reproduce, as part of their own appeal, precisely the uncertainties about agency, the rewriting of the natural as the cultural (histories of sexuality and the body, for example), and the fascination with and desire for representations that characterize (without, of course, simply equating) both consumerism and culturalism. As I have argued in terms of the panics and appeals of "physical capital," it's scarcely a matter of simply claiming immunity to the easy transitions from the economic to the erotic to the aesthetic that constitute market culture's paths of least resistance. At the same time, however, it is possible to avoid simply ratifying, by an overly hasty formalization on the level of the analysis itself, the logic of equivalence between registers "floated" in market culture.[13]

The impossibility of taking up "the culturalism question" from the outside might be epitomized by two of the neon signs designed by contemporary artist Jenny Holzer and suspended across the wall of that palace of conspicuous consumption, Las Vegas' Caesar's Palace. The first reads, MONEY CREATES TASTE, the second, PROTECT ME FROM WHAT I WANT. Whereas the first sign concisely posits something like Pierre Bourdieu's sociological reduction of taste to the laws of the market, the second couples the intimacy of the individual's desires and the intimation that they may in fact be implants.

Recent critiques of the culture of consumption have nevertheless proceeded by a panic of reduction, reducing these contradictions and choosing sides. On the one side, there is what we might call an *agoraphobic* account (the insistence on an essential opposition between the market and the self and its wants); on the other, an *agoraphilic* account (the identification of the self and its wants with an inevitable and ineluctible market). If the first or "oppositional" account acknowledges the power of the market to determine internal and affective states but opposes or dislikes it, the second or "post-oppositional" account posits an identity between the market and internal or affective states that makes opposition, or liking or disliking, simply irrelevant, since both likes and dislikes are themselves "products" of the market within. (Both sides, it may be noted, rely on a reduction of the notion of the *agora*—at once public space and market place—to a narrowly economistic status.) Not surprisingly, the first of these positions has been associated with the oppositional politics of "cultural critique," the second with the consensus formation or conservation of what has come to be called "cultural work." If the first conserves a certain paranoia about the infiltration of the market in persons, the second idealizes the identity between the market and interior states, projecting an abstract agency of "the market" beyond the vicissitudes of mere likes and dislikes.[14] The new historicist account of the market in effect replaces the Marxist appeal to the economic in the final instance with an appeal to the market in every instance.

One of the most evident effects of this idealization has been the positing of a tautological relation between individual and culture and one of the most evident effects of such a tautology has been the elision of politics as such. In his recent account of *The Body and Society,* for instance, the sociologist Bryan Turner proceeds by reading individual "disorders" as "cultural indications": in his study, as he expresses it, "In particular the disorders of women—hysteria, anorexia, and agoraphobia—are considered as disorders of society."[15] Hysteria, anorexia, and agoraphobia have, of course, become the fashionable diseases of some versions of the new historicism (particularly the new historicism of the market), and this is at least in part because the tendency has been to replace an identification of women with the natural with an identification of women with the social. In either case, what is posited is a transparently tautological relation between the individual and her material (natural or social) conditions—in this case, the transparency of the female body as leading economic indicator.

Such a reading of persons as indexes of the social reproduces what I have called the "realist tautology"—the circular relation between interior states and material conditions (between psychology and sociology) that the realist account of the individual as "socially constructed" entails.[16] The rehearsal of the realist tautology in recent cultural criticism operates by recourse to large abstractions of "the market," "property," and "possession," that, I have suggested, at once register and hypostatize the logic of equivalence that underwrites market culture. The problem, however, resides not so much in an insufficiently detailed or insufficiently periodized notion of "the market" but centrally in the principle of identity by which the market, however characterized, and self, however characterized, are positioned as the pseudo-alternatives of a tautology.

This is achieved, above all, by recourse to a highly abstract notion of desire: to the positing of desire as essentially an "equals" sign between the self and its social or cultural possibilities. This strictly culturalist notion of desire is concisely summarized by Edward Gibbon Wakefield, an early theorist of market expansion, who argued in 1833 that "According to the power of exchanging are the desires of individuals and societies."[17] The logic of sheer culturalism posits such a systemic adequation of desires, individuals, and societies. By this logic, desire becomes merely, as Joan Copjec observes, "the subjective synonym of the objective fact of the subject's construction" and thus "pleasure becomes a redundant concept and the need to theorize it is largely extinguished."[18] The project of "accounting for" persons and their wants, like the project of "accounting for" artifacts, thus has the effect of reducing persons, wants, and representations to their material condition or material conditions. One way of describing the culturalist impasse would be to say that the possibility of a politics, or at least of a cultural politics, seems, by this logic, to require precisely the discontinuity or discrepancy between persons and conditions that the program of cultural critique programmatically forecloses. Another way of saying it is that the possibility of a politics or a cultural politics seems to require the gratuitousness or discontinuity that is taken to define the "modernist" aesthetic, or, at the least, what I have been calling the aesthetics of consumption. The project of "accounting for" persons, wants, or representations collapses just that irreducible gratuitousness that, in the culture of consumption, makes up persons, wants, and the desire for representations.

Stated somewhat differently, there is no tendency more marked in recent criticism than the understanding of personhood or agency in terms of a principle of scarcity or crisis. The project of accounting for persons functions as a way of distinguishing between what counts as a person or agent and what doesn't—a way of distinguishing between persons and things that look like persons (that is, between real persons and facsimile ones.) "Mass production," the department store magnate Edward Filene observed in 1919, "demands the education of the masses; the masses must learn to behave like human beings in a mass production world. . . . They must achieve, not mere literacy, but culture."[19] The conversion of the masses into human beings (the mass production of individuals in mass culture) here links the desire for personhood with the demand for consumption and the demand for consumption with the demand for culture itself. It is the articulation of these links that Davis's *Life in the Iron Mills,* a half century earlier, begins to imagine and it is to that story of the remaking of the masses into individuals through an aesthetics of consumption that I now want to turn.

The Girl-Man

Rebecca Harding Davis's brief novel *Life in the Iron Mills; Or, The Korl-Woman* (1861) opens with these questions:

> Is this the end?
> O Life, as futile, then, as frail!
> What hope of answer of redress?[20]

The "terrible" question is whether the end of life might be no more than the end of life, whether there might be no more to Life than life. The question, the story quickly makes clear, is whether the circuit of life can be narrowed to the mere living from hand to mouth or to the narrow economy of working to eat and eating to work, to a perfected version of a "physiocrat" economy reduced to nothing but the needs of the laboring body. This is the narrowed economy or bioeconomics that is all but the state of life in the iron mills.

Davis's representation of life in the iron mills proceeds by taking up the problem of the body and its representations in its most basic terms—in terms of the fundamental relation between the hand and the head:

> "If I had the making of men, these men who do the lowest part of the world's work should be machines,—nothing more,—hands. It would be kindness. God help them! What are taste, reason, to creatures who must live such lives as that?" He pointed to Deborah, sleeping on an ash-heap. "So many nerves to sting them into pain. What if God had put your brain, with all its agony of touch, into your fingers, and bid you work and strike with that?" (34)

This passage occurs midway through *Life in the Iron Mills*. The speaker is Kirby, one of the mill owner's sons. The immediate context is the discovery of a "white figure of a woman" (31) that has been carved from korl, the "light, porous substance, of a delicate, waxen, flesh-colored tinge" (24) that is the refuse from the processing of the ore, and the encounter with the statue's maker, the inarticulate Welshman, Hugh Wolfe, who is a "furnace-hand" in the mills. The encounter produces the simple plot of the story: the theft of money by the deformed woman, Deborah, from one of the men who has accompanied Kirby on a tour of the mills; her giving of the stolen money to Wolfe, to make possible his escape from the mills and realization of his artistic ambitions; and, finally, Wolfe's almost immediate and inevitable capture, sentencing, and suicide in prison. The desire to transcend life in the iron mills, to transcend it by representing it, is thus what Kirby's fantasy of human engineering implicitly counters.

The logic of that fantasy depends upon what appears as an extension of ideology into biology, an extension by way of a literalization of the familiar notion of workers as hands. Such a tendency toward literalizing embodiment governs the general understanding of the industrial organization on the model of the organic body. One might consider here, for instance, Henry Ford's notorious classification, in his popular autobiography, of humanity as made up of a few heads and many hands, or the razor blade tycoon and utopian writer King Gillette's categorization of workers as "cogs in the machine, acting in response to the will of a corporate mind as fingers move and write at the direction of the brain."[21] (I will return to the compelling link between machine-work and writing in a moment.) That thinking and manual labor are imagined as going on in different places indicates the separation of head and hands that makes for the invention of the "suggestion box" in the later nineteenth century, a place for depositing the thoughts of hands. Not surprisingly, "Mens et manus" is also the motto of MIT, founded in 1861, the same year Davis's story of people as technology appeared. In this context, what in fact seems bizarre in Kirby's account, and one marker of the difference between midcentury corporal and later nineteenth-century corporatist accounts of the work process, is not the violence that it does to the natural unity of the body but rather the violence it does by retaining too proximate a relation to the natural body.[22] The coincidence of head and hand in a single body turns monstrous in the attempt to imagine what it would mean for a hand to have a brain.

Life in the Iron Mills attempts to make visible and make palpable—"to make . . . a real thing to you" (13–14)—the bodies set in motion by the industrial process: "the vast machinery of system by which the bodies of workmen are governed, that goes on unceasingly from year to year. The hands of each mill are divided into watches that relieve each other as regularly as the sentinels of an army. By night and day the work goes on, the unsleeping engines groan and shriek" (19). It's not hard to move from this familiar, shorthand description of the machinelikeness of persons and the personation of machines to a *Modern Times* vision of the hands of the factory system as the hands of the watch itself. Davis's account of life in the iron mills registers, albeit in significantly perfunctory fashion, the disciplines of organized bodily movement that spread throughout the social body from the late eighteenth century on, disciplines centered in but not restricted to the army and the prison, the school and the factory. Davis's account thus anticipates the double process of systematization and individualization, the making of individuals as products of the system, progressively elaborated during the course of the nineteenth century. "The individual," as Foucault puts it, "is no doubt the fictitious atom of an 'ideological' representation of society; but he is also a reality fabricated by this specific technology of power that I have called 'discipline.' "[23] The fabrication of disciplinary individuals involves the differentiation, supervision, and regulation of populations and the political "anatomy" of the body; it involves the circular process of systematizing and individualizing, the techniques of classification, individuation, and representation that fashion statistical persons. These are (and I will return to this in the following chapter) the technologies that Taylorism and the programs of systematic management rationalized at the turn of the century.

But if Davis's text anticipates disciplinary individualism, this is not yet what centers its interests or the interests of market culture generally. It is, for Davis, "the bodies of workmen" (whether this refers to the workers' bodies or the collective body of workers) and not, or not quite, their identities that she imagines as governed by the system. The vast machinery of system and the watch-work supervised by what

Davis calls "the anatomical eye" (29) are not seen as making or individualizing individuals but rather as threatening to cancel individual identity and agency altogether. What Davis counterposes to the self-canceling machinery of system, and what centers her interest in representation, is an aesthetics of self-realization: a project that correlates representation and self-making.

Life in the Iron Mills insistently pressures the double reduction that Kirby's fantasy about the making of men instances: the reduction of persons to machines or to the "hands" of the industrial machine and the reduction of interior states to material or bodily conditions. One effect of that pressure is the precipitation of uncanny hesitations about bodies and identities that proliferate throughout the novel. I have in mind the uncanniness produced, for instance, when, immediately prior to the encounter scene, we are told that Kirby "looked curiously around, as if seeing the faces of his hands for the first time" (27) or the hesitation in reference registered when, just after it, Wolfe "put his hand to his head, with a puzzled, weary look. It ached, his head, with thinking" (56). What is at issue at such moments is not so much the difference between being and having a "hand" or a body but, more exactly, the ways in which a hesitated or uncertain relation between bodies or material conditions and identity becomes, precisely, the measure of having an identity or being a person.

These conditions of identity can be clarified by looking more closely at the series of analogies and chains of reduction that *Life in the Iron Mills* obsessively generates. The story opens, for example, by drawing into relation "the masses of pig-iron" dragged "through the narrow street," the "[m]asses of men, with dull, besotted faces bent to the ground," and the "massed, vile, slimy lives, like those of the torpid lizards in yonder stagnant water-butt" (11–13). The story proceeds by testing analogies between persons and things: between, for instance, the "face of the negro-like river slavishly bearing its burden day after day" and the "slow stream of human life creeping past, night and morning, to the great mills" (12). The representation of persons as indexes of their material conditions is repeated in the insistent indexicality of the narrative itself, its here and now participation in what it represents: "come right down with me,—here, into the thickest of the fog and mud and foul effluvia. . . . I want to make it a real thing to you . . . this house is the one where the Wolfes lived" (13–14). At the same time, however, this reduction to mass or matter is figured as an abstraction to utter typicality: a story of "one of the men . . . in one of the mills . . . one rainy night," one of "myriads" whose lives were "like those of their class" and like "these their duplicates swarming the streets today" (14–15). But neither the reduction to utter materiality nor the abstraction to utter typicality—and it may be seen that these two processes are, in Davis's account, versions of the same process—gets at the immanent tension between the material body and identity, and between the typical and the individual, that makes up persons in the novel.

This tension is perhaps clearest in the remarkable description that introduces the furnace-hand/sculptor Hugh Wolfe:

> Physically, Nature had promised the man but little. He had already lost the strength and instinct vigor of a man, his muscles were thin, his nerves weak, his face (a meek, woman's face) haggard, yellow with consumption. In the mill he was known as one of the girl-men: "Molly Wolfe" was his *sobriquet.* (24)

What seems central here is not exactly the cross-gendering of Wolfe but his hesitated or hybrid identity: that is, the manner in which such a cross-naming and cross-gendering makes it impossible merely to derive identity from the natural body or merely to separate identity from the body. Hugh Wolfe's hybrid form, male and female at once, is floated in relation to the hybrid form that makes this "one" a "symptom of the disease of their class" (23). Wolfe's gender and Wolfe's class are thus rewritten in terms of a paradoxical relation between the body and its representations. In his crossing of the natural and unnatural, Hugh Wolfe is a sort of living pun ("you" "wolf"). And the structure of the pun makes graphic his hybrid form, at once material and irreducible to material, at once body and soul.

The reduction or irreduction of identity to material and the material body is tested through a chain of analogies, linking or interrupting the links between material conditions, bodies, and souls. The story insistently pressures, for example, the status of the analogies and relays between Wolfe's "squalid daily life, the brutal coarseness eating into his brain, as the ashes into his skin: before, these things had been a dull aching into his consciousness; tonight, they were reality." Or again: "He gripped the filthy red shirt that clung, stiff with soot, about him, and tore it savagely from his arm. The flesh beneath was muddy with grease and ashes,—and the heart beneath that! And the soul? God knows" (40). If *Life in the Iron Mills,* that is, insists on its indexical relation to the material conditions that it represents (that it is *part of* what it represents), it insists at the same time on an internal and irreducible, interrogative or critical distance from those material conditions (that it is not merely part of but also *about* what it represents).

The story thus implicitly invokes two different models of representation: an indexical model, by which representations are part of what they represent, and an iconic model, by which representations look like or are "about" what they represent. These are the two models that C.S. Peirce discloses, for example, in his account of the paradoxical or composite status of the photograph as an instance both of the iconic and the indexical sign. As Peirce puts it: "Photographs . . . are in certain respects exactly like the objects they represent. But this resemblance is due to the photographs having been produced under such circumstances that they were physically forced to correspond point by point to nature. In that respect, then, they belong to the second class of signs, those by physical connection."[24] And if the photograph is the exemplary "realist" form of representation, this is perhaps to suggest that the realism-effect, of *Life in the Iron Mills,* for instance, inheres in this double logic of physical connection, on the one side, and resemblance or likeness, on the other.

On this view, the uncertainty as to how realist representations are "produced," and as to the circumstances or physical forces that produce them, is bound up with the uncertainty as to whether representations are about what they represent or part of what they represent. And it's not difficult to see how this uncertainty is rehearsed in the familiar debates, across a range of recent accounts of American culture, as to whether literary texts are *about* their culture or merely *part of* their culture.

The ostensible sides of this debate are not hard to detect. What seems to be at issue is something like the chain of reduction, running from social conditions to bodies to brains to minds, that Davis tests out in Wolfe's case. Or, in the case of cultural criticism, the possibility of reducing representations to conditions (the circumstances to which they were physically forced to correspond) or the possibility of representations reflecting on, breaking with, or transcending conditions. Seen this way, we might say that oppositional criticism chooses the iconic side of the sign and what I earlier called post-oppositional criticism the indexical side. Extending the logic of these choices, whereas oppositional criticism finds in self-reflection the possibility of self-transcendence, post-oppositional criticism finds in the impossibility of self-transcendence the inevitability of representations being merely part of the culture they represent (that is, represent by epitomizing—and by epitomizing, in these cases, the inevitability of liberal market culture). And whereas oppositional criticism endorses a virtually automatic equation of self-reflection and self-transcendence, post-oppositional criticism makes the same equation and therefore moves to sentimentalize and dismiss (dismiss by sentimentalizing) both self-reflection and self-transcendence.[25] Hence both positions rehearse a logic, or panic, of reduction. For the first, invoking self-reflection functions as something of a "hyper-space" button, a way of escaping the panic of this impasse; for the second, the logic of the impasse is simply what it means to be part of a culture.

What seems to be most fundamentally at stake in these rival positions is the possibility of action or agency. Put simply, oppositional criticism goes long on agency, whereas post-oppositional criticism goes short on it. But just as both positions hold to the same reductive conflation of self-reflection and self-transcendence, both hold to the same reductive logic of action. The alternatives would seem to be, again, either a discontinuous or extra-contextual relation of actions to conditions or contexts or, alternatively, the reduction of actions to the conditions or contexts that "produce" them: either, that is, the saving of actions as breaking with or transcending conditions or the reduction of actions to reflexes of conditions. By this paradoxical logic (and here I refer back to the discussions of action and intention in the preceding chapters), actions only count as actions if they are situated but situating actions threatens to reduce them to a mere reflex of their situation.

But the problem with what George Eliot calls this "logic of human action" may by now be clear enough. As the systems theorist Niklas Luhmann neatly summarizes it: "Logically, actions are always unfounded actions and decisions are decisions exactly because they contain an unavoidable moment of arbitrariness and unpredictability. But this does not lead into lethal consequences . . . [acting] does not stop in the face of logical contradictions. It jumps, provided only that possibilities of further communications are close enough at hand."[26] The very notion of decision (*decidere*) involves such a breaking off. But it is exactly the tendency to read such a "jump" as having lethal consequences (lethal as the "aporetic" moment that suspends the possibility or predictability of action in any "full" sense or, alternatively, lethal as the "full" contextualizing that stops the possibility of unpredictability and hence the possibility of action)—it is exactly such readings that conserve a pure logic of action. Actions thus require the failure of the logic of action. But this paradox is neither lethal in theory nor lethal in practice: it is instead lethal to the theory/practice opposition.[27]

The immanent or mixed or impure account of action I have been setting out here is perhaps most striking, for our purposes, in terms of the action of representing or writing. Here again, Davis's *Life in the Iron Mills,* in its testing out of the uncertain relations between heads and hands, powerfully instances what the action of representing or writing looks like. In Davis's writing, that action holds visible the hybrid form, at once material and irreducible to material, obstrusive in the story's pressing verbal and visual puns. This double form is visible, for example, in the representation of "life in the iron mills" in the figure of "the korl-woman" (the "alternatives" floated in relation to each other from Davis's title on). It is "an eager, wolfish face" that "looks out, with its thwarted life, its mighty hunger, its unfinished work" (64) from Wolfe's projected self-representation in the figure of the korl-woman. That wolf-like figure is, clearly enough, the woman-boy counterpart of Wolfe's girl-man: "a nude woman's form, muscular, grown coarse with labor, the powerful limbs instinct with some one poignant longing. One idea: there it was in the tense, rigid muscles, the clutching hands, the wild eager face, like that of a starving wolf's" (32). The starving figure that has "no sign of starvation to the body" but is rather the victim of a "soul-starvation," hungry but "not hungry for meat" (33): the korl-woman is of course a figure for *consumption* itself. Which is not simply to pun into synonymy consumption in its bodily senses with consumption in its economic senses, but to suggest instead that the sieve-like notion of consumption is itself structured like a pun, propped on the body but not reducible to it.

And, beyond that, the "unfinished work" figured by the korl-woman is inseparable from the work of and action of representation itself. What remains, as the final image of the novel, is "this figure of the mill-woman cut in korl . . . there are about it touches, grand sweeps of outline, that show a master's hand" (64). The master's hand is revealed in the "touches" and "outline" of the *un*finished work. The master's hand, and we will return to this in a moment, is revealed in the something beyond appearance that seems to give the appearance: the work of the hand "through which the spirit of the dead korl-cutter looks out" (64). The final image of the novel is thus the necessarily unfinished work of representing the action of representation itself, representing the work with the hands that brings inscriptions to the eyes: "The deep of the night is passing while I write" (64). The hand of the writer, like the master's hand, is neither reducible to matter or the material body nor separable from it. The paradoxical economy of the body and its representations, the labor problem of hand and head, appears here as the paradox of the work of inscribing and writing itself. What is done by the hand is read by the eye: this crossing of senses draws into relation the senses of consumption, and the aesthetics of consumption, we have been tracing.

Still Lives

The privilege of relative disembodiment in consumption is not quite a living beyond the body but rather the body's aestheticization:

the body as artifact. There is, Davis writes, an "*impalpable* atmosphere belonging to the thoroughbred gentleman" (29, my emphasis). But what most strikingly belongs to the privileged body is its radical and aesthetic formality: "Wolfe caught with a quick pleasure the contour of the white hand, the blood-glow of a red ring he wore. His voice, too, touched him like music. . . ." (29). The aestheticized body that belongs to the gentlemen reappears in terms of what the psychoanalyst Jacques Lacan calls "this *belong to me* aspect of representations, so reminiscent of property."[28] After Wolfe takes the stolen money in his hand, for instance, "clutching it, as if the tightness of his hold would strengthen his sense of possession," he experiences an extraordinary vision of the artifactuality of nature: "soft floods of color in the crimson and purple flames," and "waves of billowy silver veined with blood scarlet"—in all, a "world of Beauty," and "golden mists" made "strangely real" (47–48). The lifting of the body into aesthetic representation appears as a mingling of blood and veins and silver and gold, such that bodies and artifacts and exchange concisely indicate each other. More exactly, the floods and waves of "silver veined with blood scarlet" and the "golden mists" made "strangely real" are reminiscent of the strange reality of the money form and its flows. Money is, as the Marxist philosopher Alfred Sohn-Rethel expresses it, "an abstract thing—a paradox in itself."[29]

I have earlier described the paradoxical form of "physical capital": the nexus of self-representations, representations of the body, and monetary representations in market culture. The linking of possession and self-realization in "the market" (the affective economy of possessive individualism) should by now be familiar enough. Here I want to consider in a bit more detail both the belong-to-me aspect of representations and the strange reality, at once abstract and material, that seems to confer possession and self-possession. I want, in the last part of this chapter, to reconsider the reminiscences of possession in representation: how such reminiscences make the conflation of bodies, artifacts, and money appealing and compelling in market culture; the particular forms of representation that carry the weight of those appeals and compulsions; and the vicissitudes of agency and action such representations entail.

The scene of "the market" in *Life in the Iron Mills*—and the market in the novel is emphatically represented *as* a scene—is at once thickly material and highly abstract. The strange reality of the market scene brings into relation visible and deeply embodied bodies and the abstraction of the exchange process. The market works like a working model, operating by way of the paradoxical linking of the physical, on the one side, and the formal or geometrical, on the other. The scene is viewed from and framed by the window of Wolfe's jail cell:

> It was market-day. The narrow window of the jail looked down directly on the carts and wagons drawn up in a long line, where they had unloaded. He could see, too, and hear distinctly, the clink of money as it changed hands, the busy crowd of whites and blacks shoving, pushing one another, and the chaffering and swearing at the stalls. (54)

The process of circulation takes on the sensory appeals of ear, eye, and hand (of hearing and seeing and touching) that "made the whole real" even as the jostling bodies approach the chess game abstractness of "whites and blacks" in motion. This is, in brief, the physics of bodies in motion obedient to the laws of the market, on the model of the classical physics of moving bodies, that Hobbes, adapting Newton, began to represent.[30]

There are, however, two more detailed pictures of the market that immediately follow the initial description of "the busy crowd of whites and blacks." Each complicates and more precisely locates the view of the market as an abstract thing, and together they provide an extraordinary picture of what the market looks like. The first renders the marketplace in terms that make utterly explicit what I have been describing as an aesthetics of consumption and that make utterly graphic the genre-fiction of the body and its representation proper to the market:

> How clear the light fell on that stall in front of the market! and how like a picture it was, the dark-green heaps of corn, and the crimson beets, and golden melons! There was another with game: how the light flickered on that pheasant's breast, with the purplish blood dripping over the brown feathers! (55)

The representation of the market in the genre of the still life, or better, the representation of the still life of the market itself, could not be more clearly marked. There is certainly nothing atypical about such a representation of the market. One might think here, for example, of Zola's accounts of the Paris markets—"*le ventre de Paris*"—as "*des natures mortes colossales, des tableaux extraordinaires,*" or as "*les gigantesques natures mortes des huits pavillons.*"[31] The market thus appears as the *nature morte* come to life and at least life-size. (Here it is as if the effects of colossal size and gigantism measure the shifts in scale from representations on the small fields of canvas or paper to their realization at large in the world. This intimates not merely the large effects of representation and personation created by small movements of writer's pen or painter's brush but the enlargements of possession measured by the enlargements of the body itself. "Par extension la bourgeoisie digérant, ruminant": the gigantic bodies in naturalist fiction [the bodies in the fiction of Zola—*Le Ventre de Paris* or *La Bête humaine,* for example—or Frank Norris's conflations of dentistry and mining in *McTeague,* for example] are the representations of "nature" in the form of a colossal body: the unnatural body that remakes the natural world by representing it and takes in the world by eating it.)[32]

In the realist and room-size genre of the still life, it is at least in part the arrangement and composition of the natural that makes for the appeal of these pictures of food. The still life, as Meyer Schapiro observes, conveys "man's sense of his power over things in making or utilizing them," celebrating the use and enjoyment of everyday objects, within arm's reach, as "perfectly submissive things" and as possessions: "they are instruments as well as products of his skills, his thoughts and appetites."[33] By this view, the appetite for representations—the gratification of the hungry eye—recovers a sense of agency through the representation of the effects of human agency and of the objects that serve as its representatives. The still life thus necessarily excludes the human subject and the human body since it is precisely the human subject and the human body to which the still life at every point makes reference and pays homage. One reason why we like looking at pictures of food, one reason for the bowl of fruit on the middle-class dining table or its representation on the wall across from that table, involves then the rewriting of the natural as the cultural, the aesthetic rewriting of the body and its needs. The uncertain relation between the body and its representations is converted into a reaffirmation of the body and its representations as possessions, as the agents of the possessive individual whose agency is neither separable from nor reducible to natural bodies and material possessions.[34]

The notion of the still life as the form of representation appropriate to the bourgeois style of appropriation and self-possession is of course by now a commonplace, albeit a somewhat underspecified one.[35] What is more specifically and more anxiously at issue in such a representation of the market in Davis's story, and in the ekphrastic representations of the realist novel generally, can perhaps be taken a step further by considering briefly the third and final picture of the market in this series. Wolfe, rasping at the cell bars with a "dull old bit of tin, not fit to cut korl with . . . looked out of the window again":

> People were leaving the market now. A tall mulatto girl, following her mistress, her basket on her head, crossed the street just below, and looked up. She was laughing; but, when she caught sight of the haggard face peering out through the bars, suddenly grew grave,

> and hurried by. A free, firm step, a clear-cut olive face, with a scarlet turban tied on one side, dark, shining eyes, and on the head the basket poised, filled with fruit and flowers, under which the scarlet turban and bright eyes looked out half-shadowed. The picture caught his eye. It was good to see a face like that. He would try to-morrow, and cut one like it. (57–58)

The "picture" of the mulatto girl that momentarily restores Wolfe's desire to represent, is not only the most explicit instance of the mixed identity of persons (the mixed identity that makes up persons) we have been tracing, but also the most explicit and most explicitly racialized version of the miscegenation of nature and culture in the story. The picture of the still life in persons is what I want now to examine a bit more closely.

The mulatto girl personates, and localizes, the panic about identity and agency that we have been examining. For one thing, this figure embodies the mingling of property and identity that Harriet Beecher Stowe, instancing the slave, called "living property." Beyond that, the figure of the mulatto girl embodies a mixing of person and property that here appears as something like the conflation of the portrait and the still life. If the difference between the portrait and the still life is the difference between representing persons and representing the things that represent persons, then the conflation of portrait and still life collapses the differences between the body and its representations and between persons and things. If the market appears as a still life in the realist aesthetics of consumption, then the conflation of person and still life is the representation of persons as objects of consumption. The mulatto girl's "clear-cut olive face," "cut" as if carved, and the eerie proximity of head and basket "filled with fruit and flowers"—a sort of Carmen Miranda moment in the story—signal in part the permeability of persons and things in the market (the assumption, by negation, of the gendered and racialized body). They signal in part also the counter-face of the privilege of relative disembodiment in market culture. Both indicate how the privilege of relative disembodiment—the aesthetics of the "white hand" of the gentleman, the aesthetics of the "white figure of a woman . . . a woman, white, of giant proportions" (31), the difference between "thick" skin and the "light, porous substance, of a delicate, waxen, flesh-colored [i.e. white] tinge" (24)—is allocated and "proportioned" in the competition for personhood. Both indicate that the precarious difference between person and thing appears here as the difference between consuming and being consumed and that the competition for personhood in the market is the choice between eating and being eaten.

The most spectacular instance of the conflation of the body and its representations in Davis's story is, of course, Hugh Wolfe's suicide. Cutting his body with the tin knife not fit to cut korl with, Wolfe, in his desire to "make himself" ("to *be* . . . other than he is" [25]), converts person into still life. The cutting of "the deeper stillness [of] the dead figure that should never move again" (61) provides a version of what Lukács, in his critique of the reifying tendencies of naturalist description, called "the transformation of an imitation of life into a still life."[36] But such a critique of the inanimating effects of the market seems merely the counterside of the understanding of the market as the scene of animation: of consumption as self-making and self-possession. This rhythm of suspending and recovering animation and agency structures the paradoxical still life of the market.

On one level, the appeal of the still life of the market, and the linking of representing and possessing that the still life makes visible, can't be separated from the appeal of illusionism or representation generally. One of the most familiar critiques of the culture of consumption, as we have seen, takes the form of a general critique of illusionism. The scandal of the mass culture of consumption, is, for commentators on both the cultural right (Lasch, Bell) and the cultural left (Lukács, and, in part, Bourdieu), the scandal of a fall into representationalism, illusionism, or simulation. What is taken to confirm "the triumph of the culture industry" is the fact that, as Adorno and Horkheimer express it, "consumers feel compelled to buy and use products even though they see through them."[37] But if the scandalous appeal of illusion has formed the basis of cultural critique from Plato's myth of the cave on, it's not hard to see that such a critique of mass culture amounts to the critique of culture and representation as such. By this view, the real scandal of emergent mass culture at the turn of the century was not so much the compulsive appeal of illusion—the attraction to products "even though they *see through them*"—but rather the democratization of the privilege of illusionism or simulation, a privilege antithetical to the very notion of "the mass" and its obligatory confinement to material needs and its function as a social index of "the real" and "choice of the necessary."[38]

But this is not quite to locate the belong-to-me aspect of representations, so reminiscent of property, that specifically inheres in the still life. There is, as we know, nothing more compelling than representations that discreetly make visible their own mechanisms. The illusion that fascinates but that, at the same time, one "sees through," exerts something like the fascination of the puppet that pulls its own strings. "What is it that attracts and satisfies us in *trompe-l'oeil?* When is it that it captures our attention and delights us?," Lacan asks in his consideration of the attractions of illusionistic representation and of "the picture" generally:

> At the moment when, by a mere shift of our gaze, we are able to realize that the representation does not move with the gaze and that it is merely a *trompe-l'oeil*. For it appears at that moment as something other than it seemed, or rather it now seems to be that something else. . . . It is because the picture is the appearance that says it is that which gives the appearance that Plato attacks painting, as if it were an activity competing with his own.[39]

The appearance that seems to "give" the appearance thus posits a something beyond or beneath appearance: put simply, the difference between "being and its semblance" projects the subjectivity of the subject itself. The irreducibility of the representation to the material from which it is made is also the irreducibility of the gazer to the material from which he or she is made (minimally here, the irreducibility of the gaze to the eye). By this view, the subject is the representation that seems to give the representation. And the certainty of "the real" beneath appearances is also the "certainty of being a subject."[40]

The belong-to-me aspect of representations thus binds together the logic of possessive individualism and the logic of representation; and the privilege of illusionism cannot therefore be separated from the restless project of taking possession of the "certainty of being a subject": the taking possession of one's self.[41] But it is not merely the difference between the appearance and the something beneath the appearance that makes for the attraction of the still life. The still life radicalizes this general property of representation—this tension between surface and depth and, more exactly, the attraction of the beneath appearance that seems to give the appearance—by drawing it into relation to, and by drawing into view, another aspect of the representation: the aspect of the still life that is certainly the most evident, and defining, property of the genre. The still life, as Ingvar Bergstrom concisely observes, consists in "a representation of objects which lack the ability to move."[42] What then is the relation between the stillness of the still life—the suspension of motion which is also the suspension of animation—and its deep play of appearances? And how does the interplay of stillness and illusionism lend itself to the versions of possession and self-possession we have been tracing?

Life in the Iron Mills can provide, one final time, a way of specifying and locating these large questions. The marketplace in Davis's story, and in the realist novel generally, appears at once in the scene of the still life and as the site of a ceaseless motion. The "busy crowd" "shoving, pushing"; the "free, firm step" of the mulatto girl; even the dog "down in the market . . . [that] could go backwards and forwards just as he pleased": the marketplace is characterized at once by still things and by bodies in motion. It is "just a step" from the confinement and immobility in the prison cell to the market street on which Wolfe never "should put his foot" again (55–58).

The market counterposes the still life of commodities and the motion of persons, and the scandal of the still life in persons is the reduction of persons to the immobility and, therefore, the inanimation or suspended animation of the commodity itself. "There, in the marketplace and in shop windows," Sohn-Rethel observes, "things stand still. They are under the spell of one activity only, to change owners. They stand there waiting to be sold."[43] The counterposing of motion and stillness in the market serves as constant reminder or reaffirmation of the difference between persons and things.

If what attracts us in the picture is that "moment when, by a mere shift of our gaze, we are able to realize that the representation does not move with the gaze and that it is merely a *trompe-l'oeil,*" what attracts us is at least in part the reminder of the difference between what has the ability to move and what doesn't. And what has the ability to move reminds us of the difference between the appearance and the more than appearance that seems to give the appearance. The sky that "sank down" over the iron mills, Davis writes, is "flat, immovable" (11). What links together flatness and immobility is what links the conferring of depth and the conferring of mobility and what, in turn, links the conferring of depth and mobility and the conferring of personhood. The logic of reduction that cancels personhood is thus the reduction to the flat and the immovable: to what Davis calls the "nothing beneath" (15), the reduction of what is called the *under*classes to mere masses.

Hence the attraction to the still life, like the attraction to the still life of commodities, instances the perpetual recovery of the more than appearance that allows one to set oneself in motion: the reaffirmation of agency itself. Whether enacted in the stop and go of the anticipatory ritual of window-shopping or in the identification, in passing, with the still-life representations of the model or manikin, such a rhythm of suspended and recovered motion and agency, such a rhythm of reification and personification, makes up the reanimating ritual of consumption.[44]

Mechanical Prime Movers

For Hobbes, "men are self-moving and self-directing appetitive machines" and "life it selfe is but motion." For Locke, "riches consist in plenty of moovables." For Blackstone, "personal liberty . . . consists in the power of locomotion." By the logic of liberal market culture, agency may be defined by such a postulate of mobility or by what Veblen called the "imputation of propensity." Yet if the postulate of motion or propensity defines agency and life in the marketplace, it is precisely the *inadequacy* of such a postulate to account for either life or agency in machine culture that Veblen, among others, emphasizes: not the postulate of mobility but the "postulate of automatism" defines machine culture.[45] The system of machinery, as Marx expresses it,

> is set in motion by an automaton, a motive force that moves of its own accord. The automaton consists of a number of mechanical and intellectual organs, so that the workers themselves can be no more than the conscious limbs of the automaton.[46]

Or, as Marx elsewhere writes: "a system of machinery . . . constitutes in itself a vast automaton as soon as it is driven by a self-acting prime mover."[47]

If "life itself is but motion," it is the living machine of the automaton (the railroad is the centering nineteenth-century model here of the principle of locomotion) that appears as the model of life and agency. Hence for Veblen, "the disciplinary effects" of the "machine process," and the spread of "mechanical prime movers in industry" make the imputation of propensity appear as a regressive "animism of a high grade." The conflict between the discipline of the machine process and the "sensations of consumption" is, in Veblen's terms, a conflict between impersonal "mechanisms of cause and effect" and the animistic desires of "the radiant body."[48]

Such a contradictory account of the principles of action and automatism is, we have seen, insistently generated by the body-machine complex. One of the familiar ways of understanding this contradiction has been in terms of the competing demands of the forces of production and the culture of consumption. These are, it would seem, the contradictory demands of self-discipline in the mills and self-gratification in the market that lead Daniel Bell, for instance, to characterize "the cultural contradictions of capitalism" in terms of the double life of "workers who are straight by day and swingers at night."[49] There are basic problems, as we have seen, with such a counterposing of production and self-discipline, on the one side, and consumption and self-inflation, on the other: the disciplines are by no means "opposed to" individuality (discipline makes individuals), and consumption is by no means simply "for" it (there is always "something statistical in our loves"). Yet if the statistical person is something of a commuter between these twin determinations of individuality, in motion between the market and the mills, this commuting is therefore not merely a sign of contradiction. That is, if agency in modern culture appears always in the form of a crisis of agency, such a panic about agency makes for the ritualized reaffirmations of individuality and self-possession that motivate and mobilize these contradictions. The irreducibility of these conflicts and the productiveness of these conflicts are thus alternative descriptions of the same thing. And, finally, if the cultural conflicts of capitalism solicit everywhere the offices of conflict-management, and if the stress-management consultant has become one of the archetypal figures of modernity, this is to indicate that "the crisis of late capitalism" and "the power of late capitalism" are alternate descriptions of the prime movers and still lives of the market.

Part V

The Love-Master

The Anthropology of Boys

The aim of the Woodcraft movement, observed its founder Ernest Thompson Seton, is "*to make a man.*" For Seton, a co-founder, along with Baden-Powell, of the boy scouting movement at the turn of the century, the craft of making men was the antidote to anxieties about the *depletion* of agency and virility in consumer and machine culture. As Seton puts it in the first *Boy Scouts of America* handbook (1910), he began the Woodcraft movement in America "to combat the system that has turned such a large proportion of our robust, manly, self-reliant boyhood into a lot of flat-chested cigarette smokers, with shaky nerves and doubtful vitality." In this system, "*degeneracy* is the word." Hence if the scouting "movement is essentially for *recreation,*" this is to understand recreation as re-creation, as "the physical regeneration so needful for continued national existence."[1]

It's not hard to see that such an understanding depends upon a system of analogies between the individual and the national or collective body. "As it is with the individual," Theodore Roosevelt observes in his 1899 men's club speech "The Strenuous Life," "so is it with the nation." And confronting the wilderness, Roosevelt continues, regenerates "that vigorous manliness for the lack of which in a nation, as in an individual, the possession of no other qualities can possibly atone."[2] Linking together anxieties about the male natural body and the body of the nation—linking together, that is, body-building and nation-building—Seton's or Roosevelt's programs for the making of men posit not merely that the individual is something that can be made but that the male natural body and national geography are surrogate terms. The closing of the frontier, announced by Frederick Jackson Turner in 1893, apparently foreclosed the regeneration of men through "the transforming influence of the American wilderness"—a transformation that, Turner argues, makes "a new product that is American."[3] But the closing of the frontier also indicated a three-fold relocation of the making of Americans: a relocation of the topography of masculinity to the surrogate frontier of the natural body, to the newly invented national parks or "nature museums," and to the imperialist frontier that, as the cultural historian Ronald Takaki suggests, took the form of a "masculine thrust" into Asia and Latin America.[4]

What makes such a system of analogies *operational* at the turn of the century is a related, but somewhat different, conflation of the individual and collective body. "The period of Boyhood or the Gang Period corresponds racially to the tribal period," the 1914 *Handbook for Scout Masters* informs us: "The early Adolescent or Chivalry period," the handbook continues, "is racially parallel to the Feudal or Absolute Monarchial period with its chivalric virtues, vices and actions."[5] By 1914 such a correspondence of individual and racial development—the notion that ontogeny repeats phylogeny and that the biological and psychological evolution of the individual recapitulates the evolution of the race—is something of a commonplace.[6] Hence Seton's assertion that the boy is "ontogenetically and essentially a savage" or the prominent psychologist G. Stanley Hall's observation that the male adolescent's activities were analogous to imperialism, an imperialism that Hall described as "the ethnic pedagogy of adolescent races." The invention of adolescence at the turn of the century—most notably in Hall's massive and highly influential *Adolescence* (1904)—is bound up with a more general anthropology of boyhood and pedagogy for the making of boys into men.[7] Such an anthropology and such a pedagogy mobilize the *relays* between the natural and national body we have begun to sketch (See Figure 1).

In the pages that follow, I will be concerned with the set of relations that constitute what I have called the topography of masculinity in America at the turn of the century. More specifically, I will be concerned with some of the ways in which anxieties about agency, identity, and the integrity of the natural body are distributed across physical landscapes. I will be concerned, that is, with the relays between individual physiology and physical geography that make up a national geographics. Here my examples are drawn from the popular literature of adolescence and the making of men produced at the turn of the century—specifically, Stephen Crane's novel *Red Badge of Courage* (1895) and the wilderness or "wilding" stories of Jack London.[8] I want to take

Figure 1. Theodore Roosevelt: Body-building and Nation-building.

up as well an apparently opposed, but, I would suggest, fundamentally related counter-tendency in these accounts of the national body.

These pages thus form part of the series of case studies in which I have traced the ways in which the body and the machine are coordinated in late nineteenth- and early twentieth-century American culture. I have been examining throughout what the preacher Josiah Strong, in his influential *The Times and Young Man* (1901), called "the problem of *THE BODY*" in what Thorstein Veblen, among others, called "machine culture." I have argued that nothing typifies the American sense of identity more than the love of nature (nature's nation) except perhaps its love of technology (made in America). It's this double discourse of the natural and the technological that, I have traced, makes up the American "body-machine complex." And I want here to take this account of the traffic between the natural and the technological one step further by looking more closely at the "naturalist" project of making men.

Making Men

The invention of boyhood or "boyology" and adolescence is part of the reasserting of *the natural* in machine culture, and, correlatively, with a modeling of the nation on the male natural body.[9] Yet if the reassertion of the natural and of the natural body in machine culture is generally seen as contradictory or compensatory, such an account is, as may already be clear, somewhat misleading. It's misleading, in part, because it is the very notion of *what counts as natural* that is being negotiated in these practices and discourses. But it's misleading as well because the reassertion of the natural in machine culture is not finally incompatible with the anti-natural and anti-biological biases that underwrite that culture. If turn-of-the-century American culture is alternatively described as naturalist, as machine culture, and as the culture of consumption, what binds together these apparently alternative descriptions is the notion *that bodies and persons are things that can be made.*

The panic about the natural body in machine culture—the simultaneous invocation and denigration of the natural and biological in "the culture of consumption"—are perhaps clearest in the double logic of *consumption* itself (see also Parts II and IV). Stated simply, the notion of consumption depends upon a condensation or conflation of bodily and economic states, of the individual and social body. Such a conflation is made visible in the punning relation between consumption in its bodily senses (as eating or as the body's being eaten away or wasted away by disease) and in its economic senses (as waste or as want that exceeds physical need). As we have seen in the preceding chapter, one of the most evident paradoxes of the insistently paradoxical notion of a culture of consumption is the manner in which a style of life characterized by its excessiveness or gratuitousness—by its exceeding of, or disavowal of, physical and natural and bodily needs—is yet understood on the model of the natural body and its needs. The tendency to understand consumption on the model of hunger or eating—in Baudrillard's terms, "as a process of craving and pleasure" or "as an extended metaphor of the digestive functions"—might be taken to emblematize the simultaneous invocation and transcendence of the natural and bodily in the culture of consumption.[10]

Seton, for instance, understands consumption as a symptom of a perverse turning away from the natural. As he puts it in the first *Boy Scouts of America* handbook, "Consumption, the white man's plague since he has become a house race, is vanquished by the sun and air, and many ills of the mind also are forgotten, when the sufferer boldly takes to the life in tents. Half our diseases are in our minds and half in our houses."[11] Seton's account of the national disease of consumption, its etiology and cure, invokes a familiar opposition of the artificial and female indoor space of domesticity and conspicuous consumption, on the one side, to male and natural outdoor life, on the other. As Missouri Senator George G. Vest put it, in his 1883 defense of the founding of the first national park, Yellowstone would serve "as a great breathing-place for the national lungs."[12] The return to nature thus appears as the antidote to consumption conspicuous in body and nation both.

Yet if the naturalist critique of the effects of an "over-civilized" culture of consumption seems clear enough, the return to nature turns out to be a bit more complicated. "Our business," as Robert Baden-Powell expressed it, "is not merely to keep up smart 'show' troops but to pass as many boys through our character factory as we possibly can."[13] The boy scout character factories, like the "man factories" that Mark Twain's Connecticut Yankee Hank Morgan devises to turn groups of boys into brigades of workers, couple the natural body and the disciplines of machine culture.

For one thing, the scouting organization itself is something of a model of uniform, and uniformed, mass production. As David Macleod has shown, "its appealing and standardized program, its strategy of replicating small units supervised by a promotionally aggressive bureaucracy," made possible a standardized regimen in the building of the American boy. Just as the anthropology of boyhood and adolescence had reoriented the contest between adults and boys from "a few convulsive struggles for autonomy" to a series of grades and "endless little tests along a finely calibrated course," the character factory, like the man factory, standardizes the making of men, coordinating the body and the machine within a single system of regulation and production.[14]

One might instance here the coordination of bodies and standards in the rise of eugenics and euthenics, in movements of scientific motherhood and "physical culture." As the founder of the eugenics movement Francis Galton observed in his "Eugenics: Its Definition, Scope and Aim," which appeared in *Nature* in 1904: "Eugenics cooperate with the workings of nature. . . . What nature does blindly, slowly and ruthlessly, man may do violently, quickly and kindly. . . ." In order to arrest degeneracy, Galton's disciple Karl Pearson argued, it is necessary to "increase the standard, mental and physical, of parentage." One might instance as well the popularity of the body-builder Eugen Sandow, by 1890 one of the most famous athletes in the United States, and of his writings (or ghost-writings) *Body-Building, or Man in the Making* (1905) and *The Construction and Reconstruction of the Human Body* (1907). As Conan Doyle (a friend of Baden-Powell and author of imperialist as well as metropolitan fantasies) wrote in his introduction to the latter: "the man who can raise the standard of physique in any country has done something to raise all other standards as well." These technologies of regeneration, of man in the making, make visible the rewriting of the natural and of the natural body in the idiom of scientific management, systems of measurement and standardization, and the disciplines of the machine process.[15]

What then looks like, from one point of view, a return to nature looks, from another, something like the opposite, a turn against nature. But neither view, it turns out, is adequate to account for the double discourse of the body and the machine in naturalism. In the course of his discussion of the primitivist pursuits that "*make for manhood,*" Seton, for instance, proposes a new standard of evaluation for these pursuits—proposes, in fact, what he calls "*Honors by Standards*":

> The competitive principle is responsible for much that is evil. We see it rampant in our colleges to-day, where every effort is made to discover and develop a champion, while the great body of students is neglected. That is, the ones who are in need of physical development do not get it, and those who do not need it are over-developed. The result is much unsoundness of many kinds. A great deal of this would be avoided if we strove to bring all individuals up to a certain standard. In our non-competitive tests the enemies are not "*the other fellows*" but *time and space,* the forces of Nature.[16]

Seton's account of "honors by standards" enacts in miniature the rewriting of the natural body in the idiom of machine culture that concerns us here. What this passage registers is one version of the transition from *competitive individualism* and market culture to what I have been calling *disciplinary individualism* and machine culture.[17] The replacement of unregulated competition by abstract and impersonal standards is also imagined as the replacement of the individual and organic body by the collective body of the organization—here concisely registered in the bridge-making phrase "the great body of students." This is not, however, to replace individuality with standards (honors *versus* standards) but to make the achievement of the standard the measure of individuality (honors *by* standards)—to make individuals as *statistical persons*. But if statistical persons here are imagined as *un*natural or anti-natural (the competition among persons has become the competition between persons and "the forces of Nature"), the translation of the natural into abstract and disembodied laws of force ("*time and space,* the forces of Nature") points not merely to the unnaturalness of persons but to the unnaturalness of nature itself. Indeed, since persons are here personations, the personification of the natural as "Nature" indicates precisely Nature's unnaturalness.

Living Diagrams

The notion of the unnaturalness of nature has, of course, become something of an interpretive standard in recent cultural criticism. The rule-of-thumb that has guided much recent criticism might be restated in these terms: When confronted by the nature/culture opposition, choose the culture side. This criticism has thus proceeded as if the deconstruction of the traditional dichotomy of the natural and the cultural indicated merely the elimination of the first term and the inflation of the second. Rather than mapping how the relays between what counts as natural and what counts as cultural are differentially articulated, invested, and regulated, and rather than reconsidering the terms of the nature/culture antinomy and the account of agency that antinomy entails, the tendency has been to discover again and again that what seemed to be natural is in fact cultural.

What has sponsored this choice are, of course, the political benefits that seem to accrue from it: If persons and things are constructed, they could, at least in principle, be constructed differently. The choice of the culture side thus seems guided by the anxieties about choice and agency—by the melodramas of uncertain agency—that have marked a good deal of recent cultural criticism.

But as the examples I have to this point given may already suggest, this virtually automatic anti-naturalism or sheer culturalism does not come with a specific political program "hard-wired" into it. And the panic about agency—the "sublime" melodrama of agency suspended and recovered—reenacted in recent cultural criticism has something of a resemblance to the rituals of agency depletion and regeneration that, for example, "make for manhood" in naturalist discourse. The resolutely abstract account of "agency" in recent cultural work, like the resolutely abstract account of "force" that governs the emphatically male genre of naturalism, has the effect of guaranteeing the restaging of the drama of agency-in-crisis and the choice of the culture side. I am suggesting that these recent, allegedly "post-realist," accounts continue to inhabit, and to conserve, the forms and tensions of the realist body-machine complex.

More specifically, the tension between honors and standards registers what the historian of technological systems James R. Beniger has called "a crucial transition in human thought about programming and control between the 1870s and 1930s." This transition involves, in part, the shift from the "untrammelled market economy" of the mid-nineteenth century to the modern economy of systematic management. The transformation of the market by increasing systematic and administrative control may be seen as the progressive replacement of the "invisible hand" of the market economy by what Alfred Chandler describes as the "visible hand" of the managerial economy: the general achievement of standardization, programming, and processing of materials, persons, and information. The "levelling of times and places" in a culture of standardization is part of the progressive replacement of animate or naturally occurring sources of energy (animal power and water power, for instance) with inanimate sources of energy; of living or naturally occurring styles of motion by mechanisms and by "mechanical prime movers"; and, most generally, by (in Beniger's terms) "the transcendence of the information-processing capabilities of the individual organism by a much greater technological system."[18]

These transformations in the forms of energy, motion, and organization are part of the process that Thorstein Veblen, for example, advocated as the replacement of what he invidiously called "the radiant body" by "dispassionate sequences of cause and effect."[19] The shifts from animate to inanimate forms of energy and motion, and technological "transcendence" of the natural body—these transformations, I will be suggesting in the considerations of naturalist fiction that follow, make for the renegotiation of what Veblen calls the "vague and shifting" line between the animate and the inanimate in machine culture, the "discrimination between the inert and the animate" by which "a line is . . . drawn between mankind and brute creation."[20] But before returning to the manner in which Crane and London represent such a renegotiation, I want to consider briefly several instances that epitomize what such a redrawing of the line looks like.

The first is a brief fantasy that appears in Henry Ford's autobiographical *My Life and Work* (1923). The production of the Model T required 7882 distinct work operations, but, Ford noted, only 12% of these tasks—only 949 operations—required "strong, able-bodied, and practically physically perfect men." Of the remainder—and this is clearly what he sees as the major achievement of his method of production—"we found that 670 could be filled by legless men, 2,637 by one-legged men, two by armless men, 715 by one-armed men and ten by blind men."[21] If from one point of view, such a fantasy projects a violent dismemberment of the natural body and an emptying out of human agency, from another it projects a transcendence of the natural body and the extension of human agency through the forms of technology that represent it. This is precisely the double logic of prosthesis and it is also the double logic of a sheer culturalism that posits that the individual is something that can be made (see Figure 2).

The replacement of the natural body by the artificial body of the organization entails a transformation in production that is also a politics of reproduction. That is, the technologies for the making of men devised in naturalist discourse provide an anti-natural and anti-biological alternative to biological production and reproduction: the mother and the machine are, in the naturalist text, linked but rival principles of creation. These technologies of reproduction make up what I have described as "the naturalist machine" (see Part I). One form this competition between principles of production takes appears in the redefining of the category of production itself. Such a redefinition, in the naturalist discourse of force, displays in part a compensatory male response to a threatening female productivity: a compensation already implicit in such a counterposing of "male" and "female" powers or principles. It displays also the "culturalist" desire to devise an anti-natural and anti-biological countermode of making, a desire to "manage" production and reproduction.

Not surprisingly, such a desire is clearest in the work of Frederick Winslow Taylor, "the *father* of scientific management." The real achievement of Taylorization is not the invention of a system of industrial discipline. As Foucault, among others, has shown, the invention of disciplinary practices, in the army, the hospital, the factory, the school, has a long history, extending throughout the social body, from the late eighteenth century on. The real innovation of Taylorization was the redescription of managerialism and supervision in the idiom

Figure 2. "Education of the Movements of the Wounded Soldier," from *The Physiology of Industrial Organization* (1918), by Jules Amar.

of production. That is, the real innovation of Taylorization becomes visible in the incorporation of the *representation* of the work process into the work process itself—or, better, the incorporation of the representation of the work process *as* the work process itself. Taylor, in effect, rationalized rationalization.

What this amounts to is in part a system of supervision by representation: as James Howard Bridge put it in his *Inside History of the Carnegie Steel Company* (1903), "The men felt and often remarked that the eyes of the company were always on them through the books."[22] Systematic management indeed involves a coordination of production and representation: "the process of production is replicated in paper form before, as, and after it takes place in physical form."[23] But it is finally just this distance between physical processes and processes of representation that systematic management manages to eliminate.

In brief, the radical transformation in the thinking about programming and control from the 1870s to the 1930s that James Beniger has called "the control revolution," involved, above all, a rethinking of the problem of representation, communication, and information-processing: that is, the understanding of processes of representation—the always-material forms of information-processing—as production, and the understanding of production as processing, programming, and systematic communication. As the systems theorist Niklas Luhmann has recently summarized it: "The system of society consists of communications. There are no other elements, no further substance than communications."[24] One reason why Maxwell's famous sorting demon (the hypothetical being that sorted fast and slow molecules so as perpetually to maintain energy in a closed system) seemed so paradoxical, at the time, was the basic difficulty in understanding sorting—information-processing—as work. And one reason why such a paradox now seems so commonplace is the basic difficulty in understanding work as anything other than as a process of sorting, representing, or programming.[25]

One effect of such a collapsing of the distinction between production and processing is a collapsing of the distinction between the life process and the machine process. Taylor's application of the principles of scientific management to the work process (in *The Principles of Scientific Management* [1911]) closely paralleled Jacques Loeb's application of the principles of scientific engineering to the life process (in, for instance, *The Mechanistic Conception of Life* [1912]). Both processes appear as control-technologies, as self-reproducing programs: "information processing and communication, insofar as they distinguish living systems from the inorganic universe, might be said to define life itself—except for a few recent artifacts of our own species."[26] And it's precisely that exception—the invention of the technological system—that redraws the vague and shifting line between the animate and the inanimate and draws into relation a transformation in production and a politics of reproduction. For if the modern engineer-manager "does not create, but moderates and adjusts," this is because creation, production, and agency are themselves to be seen as processes of adjustment, replication, and systemic regulation.[27] These are some of the ways in which the appeal of systematic management is neither reducible to nor separable from anxieties about the gender of production and reproduction.

It should by now be clear, however, that naturalist discourse registers such a transformation in production in terms of what I have called the double logic of prosthesis: in terms, at once, of panic and of exhilaration. The discourse of naturalism is situated at the crux of this transformation: at the excruciated moment of confrontation between bodies and machines. Such a confrontation is richly emblematized in the fascination with forms of representation—with maps, diagrams, grids, models, and pictures—and with the fascinated juxtaposition of bodies and representations. We might consider, for example, the well-known photographic work of Eadweard Muybridge, with its juxtaposition of bodies in motion against gridded backgrounds (See Figure 3). Muybridge's fascination with the technological replication of "the natural" appears also in the model he was building in his backyard at the time of his death—a scale model of the Great Lakes.[28] It's such a renegotiation of bodies, technologies, and landscapes that I want next to look at, very briefly, in the work of Stephen Crane.

But such a moment of confrontation might also be emblematized by a somewhat different set of relays between representation, discipline, and the transcendence of the male natural body. The calibration of bodily movements, the conversion of bodies into living diagrams, the practices of corporeal discipline that appear at once as a violation of the natural body and its transcendence: there is something like a resemblance between the mechanisms of scientific management and the invention of sadomasochism, also at the turn of the century. As Gilles Deleuze has argued in his study of Sacher-Masoch, *Coldness and Cruelty,* one discovers in these systems of discipline and punishment "a naturalistic and mechanistic approach imbued with the mathematical spirit" and also an obsessive imperative to represent: "Everything must be stated, promised, announced, and carefully described before being accomplished." One discovers, above all, a "transmutation" of animal nature and the natural body into the human and the artifactual, a redrawing of the line between the human and brute creation, a line here drawn by the whip.[29] There is perhaps something of a resemblance

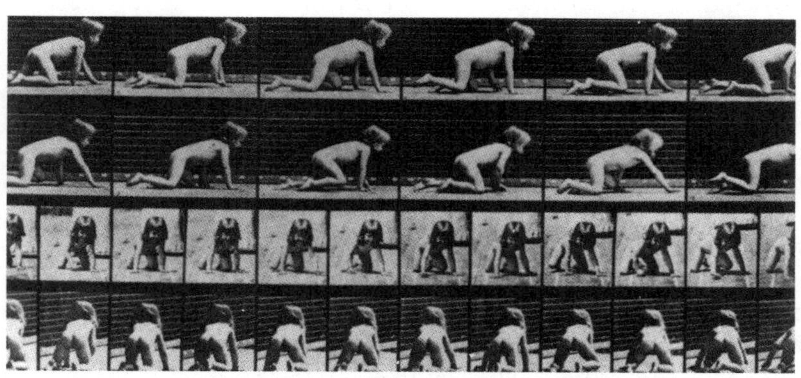

Figure 3. *Animals in Motion* (1899), Eadweard Muybridge.

between Sacher-Masoch's fantasies of discipline in such writings as *Venus in Furs, The Wolf,* and *The She-Wolf;* Seton's cub scouts, den mothers, and wolf packs; and the accounts of discipline and bondage in Jack London's stories of coldness and cruelty, stories of men in furs such as *The Sea Wolf, White Fang,* and *The Call of the Wild.* Following a brief discussion of Crane's writing, it is the pleasures of systematic management and man in the making that I will be looking at, in conclusion, by way of the work of London.

National Geographics

Perhaps the best-known American story of the anthropology of boyhood and the making of men, at the turn of the century, is Stephen Crane's *Red Badge of Courage* (1895).[30] I have earlier observed that *Red Badge* in effect tells two stories at once, a love story and a war story (see Part III). On the one side, there is an "inside" story of the "quiver of war desire" (55), of male hysteria and the renegotiation of bodily and sexual boundaries and identities. These insecurities about boundaries are registered, for instance, in the "bloody minglings" (104) that give the soldiers a "purchase on the bodies of their foes" (79), and in the "potent . . . battle brotherhood" (31) of an eroticized violence, of body-machines rhythmically "thrusting away the rejoicing body of the enemy" (101). These insecurities are registered also in the fears of unmanning and infantilization that make up this inside narrative: the threats of bodily dismemberment that are frequently localized (if that is the right word) in well-marked scenarios of castration (battlefields "peopled with short, deformed stumps" [71]) and fantasies of maternal engulfment ("as a babe being smothered" [31]). On the other side, there is an "outside" story of social discipline and mechanization, of territory taken and lost, of body counts and the industrial and corporate disarticulation of natural bodies and the production of the disciplined, collective "body of the corps."

It's not quite a matter of equating these twin stories, the psychological and the sociological. Nor is it, certainly, a matter of choosing between them. If the inside story and the outside story seem interchangeable, this is precisely because it's the boundaries between inside and outside that are violently being renegotiated, transgressed, and reaffirmed. This is what scenes of battle look like in Crane's story:

> Wild yells came from behind the walls of smoke. A sketch in gray and red dissolved into a moblike body of men who galloped like wild horses . . .

> The youth shot a swift glance along the blue ranks of the regiment. The profiles were motionless, carven, and afterward he remembered that the color sergeant was standing with his legs apart, as if he expected to be pushed to the ground.
>
> The following throng went whirling around the flank. Here and there were officers carried along on the stream like exasperated chips. . . . A mounted officer displayed the furious anger of a spoiled child. He raged with his head, his arms, and his legs. . . . The battle reflection that shone for an instant in the faces on the mad current made the youth feel that forceful hands from heaven would not have held him in place if he could have got intelligent control of his legs.
>
> There was an appalling imprint on these faces. The struggle in the smoke had pictured an exaggeration of itself on the bleached cheeks and in the eyes wild with one desire.
>
> The sight of this stampede exerted a floodlike force that seemed able to drag sticks and stones and men from the ground. They of the reserves had to hold on. They grew pale and firm, and red and quaking. (28–29)

The transgression of boundaries involves not merely the bloody mingling of bodies of individuals and the "moblike body of men" but also the "floodlike force" merging bodies and landscapes (assaulted "flanks," and, elsewhere, forming fronts and protecting rears). The "dissolving" of men into artifacts (sketches and imprints, "motionless, carven") is also the uncertain relation between surface "reflection" and interior states: the "fitting" of inside and outside in the struggles of individual and regimental bodies to pull themselves together ("*in* the faces *on* the mad current," "*in* the smoke . . . *on* the bleached cheeks"). The fit of bodily and group identities is signaled by redundancy or tautology ("They of the reserves had to hold on") and by the erotics of a body of men "wild with one desire" ("They grew pale and firm, and red and quaking").

The becoming-artifactual of persons, in these descriptions, is perfectly compatible with the substitution of the regimental and regimented body for the natural body—the military "making of men." And the "drilling and training" that makes men into members, components of the war "machine," also substitutes the invulnerable and artificial skin of the uniform-armor for the vulnerable and torn natural body ("He held the wounded member carefully away from his side so that the blood would not drip upon his trousers" [28]).

These primal scenes of battle are, finally, struggles to make interior states visible: to gain knowledge of and mastery over bodies and interiors by tearing them open to view. This is what the "shock of contact" (103) with the male natural body and between male bodies looks like, as it is enacted in the not socially unacceptable context of battle.

There is a good deal more that might be said about scenes such as these. For now, it may be noted that if anxieties about identity and virility are distributed across natural landscapes; if it seems possible to work out the transgression of bodily boundaries through the transgression of geographical boundaries; if it seems possible to restore or subtract agency or manhood through the annexing or subtraction of pieces of territory—then it's precisely the coordination of interior states and exterior and territorial states that makes such excruciated crossings thinkable. "The terrain of their rage," as Klaus Theweleit observes, albeit of a different body of men, "is always at the same time their own body."[31]

From one point of view, the description of such social processes in terms of natural processes appears as a form of *naturalization* of the social. But to the extent that the notion of naturalization has become virtually synonymous with mystification or disguise, such a notion is inadequate to describe these actions—actions which are at once identified with *and realized through* the reworking of natural processes and landscapes. That is, the channeling of "floodlike forces" describes at once the regulation of bodily flows and identities and the work of civil engineering. I have in mind here the range of turn-of-the-century work that includes, for instance, the culture-work of channeling, bridge-building, and canalization ("the roads . . . were growing from long troughs of liquid mud to proper thoroughfares" [5]); the strenuous exploit of Theodore Roosevelt and the building of the Panama Canal ("The land divided, the world united"); the emergence of the civil engineer as culture hero in the literature of the 1890s (in the novels of Richard Harding Davis, for instance). Freud's model for the mechanisms of sublimation that extend the culture-work of the ego into the territory of the id is, not surprisingly, the draining of the Zuider Zee.[32] Put simply, to the extent that the anti-biological and anti-natural biases of naturalism involve, as we have seen, the transcendence of "the natural" and "the female" both, they involve the transcendence of a female/nature, identified with liquid interiors and flows. Such a channeling of natural floods into orderly movements thus forms part of the technologies for the making of men we have been tracing here (See Figure 4).

It might even be said that the processes of man-in-the-making assume something like the form of a continuous "flow technology." The effect of drilling and training ("he was drilled and drilled and reviewed, and drilled and drilled and reviewed" [10]) is to transform

Figure 4. "I Took the Canal Zone." Theodore Roosevelt at the Culebra Cut, Panama, 1906.

interior states, such as seeing, thinking, planning, and feeling, into visible and measurable movements of the body. (That is, to transform interior states into *esprit de corps.*) The disciplines of organized bodily movement that make up systematic management are made visible not merely in the military corps or in the time-motion studies of Taylor and the Gilbreths but, more generally, in the identification of the life process and the work process with the imperative of keeping things and bodies in directed motion.

This is in part because "physical movement, processes, and speed, present the most pressing problems of control" for industrial production. But it is also because the dream of directed and nonstop flow forms part of the psychotopography of machine culture. As the historian Daniel Boorstin puts it, "If Ford had succeeded perfectly, a piece of iron would never have stopped moving, from the moment it was mined until it appeared in the dealer's showroom as part of a completed car"—not that, if he had succeeded perfectly, the automobile would quite stop there, in the still life of the display room. "The watchword, then, was 'Flow!' "[33]

Or, as Jack London expresses it, "life is movement; and the Wild aims always to destroy movement":

> It freezes the water to prevent it from running to the sea; it drives the sap out of the trees until they are frozen to their mighty hearts; and most ferociously and terribly of all does the Wild harry and crush into submission man—man, who is the most restless of life, ever in revolt against the dictum that all movements must in the end come to the cessation of movement.

Hence the mastery of men—the *un*natural mastery that revolts against the natural law of "the Wild" that freezes motion—consists in technologies of flow and continuous motion. The mastery of men consists in "their capacity to communicate motion to unmoving things; their capacity to change the very face of the world."[34]

Men in Furs

There is perhaps no more compelling exploration of the cluster of relations I have been examining here than in the work of a writer who seems as far from the concerns of either machine culture or the culture of consumption as one could possibly be. But the violent confrontations between the natural and the cultural, between the wants of the body and the disciplines of correct training, and, most fundamentally, between the life of motion and the threat of a cessation of life and motion—all are nowhere more powerfully represented than in the writings of Jack London. London's work reduces these conflicts to their most rudimentary forms: the culture of consumption to "the law of the meat" and the disciplines of machine culture to "the toil of trace and trail." The difference between life and death in London is the difference between the body staying in motion and the freezing of bodies and motion. And the difference between what counts as an agent and what doesn't is the difference between eating and being eaten. London's writings enact both the fascinations of the body-machine complex ("Body and brain, his was a more perfected mechanism" [*WF* 203]) and the violence registered in such crossings or *miscegenation of nature and culture*—a miscegenation of the natural and the cultural most apparent, of course, in the figure of the wolf-dog and other men in furs.

The abstract and zero-degree white zones that London's persons, or personations, move across are governed by the dispassionate laws of force that London, in *John Barleycorn* (1913), calls the "White Logic": "the antithesis of life, cruel and bleak as interstellar space, pulseless and frozen as absolute zero" (*JB* 1094).[35] The abstract and emblematic character of this landscape makes explicit the links between bodies and territories, and the struggle to stabilize the uncertain relations between inner and outer states: the struggle, for example, of men who are "all equally as mad to get to the Outside as they had been originally to get to the Inside" (*WF* 253). It makes explicit also the paradoxical economy of London's call of the wild, what I have been describing as the unnaturalness of Nature in naturalism. That is, if "the Wild," and its White Logic, are "the antithesis of life" (the enemy of motion), this is to indicate the unnatural or "beyond the natural" (*WF* 172) character of life (motion) itself. Such a turning away from the natural makes for what might be described as the compulsory unnaturalness or compulsory perversity of naturalist discourse. This perversity is revealed, on the one side, in the unnatural disciplines of the machine process and, on the other, in the unnatural disciplines of naturalist sexuality.

The twin principles of gold and the machine are the economic principles that put bodies in motion across the landscape of the great white male North: "Because men, groping in the Arctic darkness, had found a yellow metal, and because steamship and transportation companies were booming the find, thousands of men were rushing into the Northland" (*CW* 21). The delivery of men here indicates something more than transportation-technologies. It registers as well what I have elsewhere called the perverse *accouchements* of the naturalist text: the anti-biological and technological making of men and the replacement of the mother by the machine. If, as London puts it in *White Fang* (1906), "the white men came from off these steamers" (204–05), this is because the mechanical reproduction of men (that is, white men) is explicitly counterposed to the biological reproduction of persons.

Hence London, in *The Sea-Wolf* (1904), describes Wolf Larsen's men in these terms: "They are a company of celibates, grinding harshly against one another. . . . It seems to me impossible that they ever had mothers . . . they are hatched out by the sun like turtle eggs, or receive life in some such similar and sordid fashion." These celibate machines are "a race apart, wherein there is no such thing as sex" (*SW* 583). Hence Baden-Powell, in *Young Knights of the Empire* (1916), describes the engine room of the steamship *Orsova* in these terms:

> And it is indeed an impressive sight to stand below these great monsters of steel and watch them faithfully and untiringly pounding out their work, all in order and exactly in agreement with each other, taking no notice of night or day, or storm or calm, but slinging along at all times, doing their duty with an energetic goodwill which makes them seem almost human—almost like gigantic Boy Scouts.[36]

The mechanical process of producing men is thus a process of systematic management—the formation of the disciplinary individual. And the system of disciplinary individualism involves not merely the individualization and specialization of work and workers (the "division of labor" and "special knowledge" that London takes up, for example, at the opening of *The Sea-Wolf*). It involves also the Taylorization of bodies and interiors: what London calls the "achieve[ment of] an internal as well as external economy" (*CW* 25). It involves, most fundamentally, the identification of the life process and the machine process, the "coordination" of the body and the machine.[37]

Not merely does the toil of trace and trail transform "sullen brutes" into ideal workers—"straining, eager, ambitious creatures" (CW 33). ("So he worked hard, learned discipline, and was obedient" [WF 196].) Bodily processes are identified with efficient machine processes, internal and external economies all in order and precisely coordinated.[38] For London, as for Seton, what this means, finally, is the bringing of individuals up to efficient standards through a mastery of the laws of "time and space, the forces of Nature."[39] London's accounts of the wild often resemble time/motion studies, and "the sounding of the call" appears as a "time-card . . . drawn on the limitless future" (CW 73). This is the unnatural Nature that Veblen neatly condenses in his notion of "the instinct of workmanship."

But the disciplines of Systematic Management are bound up with another form of S/M in the Klondike. One achievement of disciplinary individualism is, we have seen, the transformation of interior states and natural bodies into supervisable and finely calibrated spatial movements. Another and correlative technique is the rigorous segregation or partitioning of working bodies, spaces, times, and functions: the rank-and-file of bodies side by side, without promiscuous contact. One effect of these technologies is to rule out what Crane calls "the shock of contact"—the flooding of physical and individual boundaries. Or, as London writes of one of his masters of time and space: "He could not endure a prolonged contact with another body. It smacked of danger. It made him frantic" (WF 202).

Yet if these figures display again and again "the panicky impulses to avoid contact" (WF 248), they display as well the desire for the "smack" of contact. Multiplying at once the system of differences and (therefore) the possibilities of transgression, systematic management makes possible the hidden desires for a contact at once proximate and proscribed, a contact eroticized and rendered violent by this combination of proximity and proscription. Not surprisingly, the workplace, in naturalist discourse, is an arena at once of discipline and of eroticism, of segregation and of a promiscuous mixing, across sexual, class, and rank boundaries.[40]

Such an erotics of discipline is nowhere clearer than in London's stories of discipline and bonding, of coldness and of cruelty. The counter-side of the fear of contact is "the deliberate act of putting himself into a position of hopeless helplessness"—the initiate's self-surrender to "the love-master." He was, London observes, "in the process of finding himself":

> There was a burgeoning within him of strange feelings and unwonted impulses. . . . In the past he had liked comfort and surcease from pain, disliked discomfort and pain. . . . But now it was different. Because of this new feeling within him, he ofttimes elected discomfort and pain for the sake of his god. . . . It was an expression of perfect confidence, of absolute self-surrender, as though he said: "I put myself into thy hands. Work thou thy will with me." (*WF* 244, 248)

Learning to love pain and the god-like hand of his master, White Fang learns to love at once the pleasure of unnatural acts (acts contrary to every "mandate of his instinct") and the pain of turning from "the natural" to "the cultural." And since the notion of a turning away or perversion from the natural to the cultural can scarcely be separated from modern sexuality "as such," what White Fang learns in learning to love pain are the culturalist possibilities of love "as such."[41]

Which is not to say that London merely represents the compulsory perversity of naturalist sexuality in the disguised and acceptable form of the animal fable. The "naturalization" of the disciplines of machine culture is, I have argued, inseparable from the redrawing of the uncertain line between the human and the animal, between "mankind" and "brute creation." Along the same lines, London's stories of men in furs make utterly explicit what I have been describing as the transcendence of the natural body in the naturalist project of making men. I want to close by taking up, very briefly, a somewhat different implication of such stories of men in furs and of what might be called naturalist *skin games* generally.

Writing in 1859 about the still-recent technological achievement of photography, Oliver Wendell Holmes described that achievement in these startling terms:

> Form is henceforth divorced from matter. . . . We have got the fruit of creation now, and need not trouble ourselves with the core. Every conceivable object of Nature and Art will soon scale off its surface for us. Men will hunt all curious, beautiful, grand objects, as they hunt cattle in South America, for their skins and leave the carcasses as of little worth.[42]

It might be suggested that if photography is the realist form of representation par excellence, taxidermy is the form of representation proper to naturalism. There is something of a continuity between Holmes's celebration of the ruthlessly superficial hunting and skinning with a camera and the dioramas of stilled life that make up, for example, the visual communion between "man" and "nature" in the Roosevelt Memorial of the American Museum of Natural History.[43] And there is something of a continuity between these exhibitions of typical Nature captured by the naturalist art of taxidermy—visual displays that seem to hover midway between the *tableau vivant* and the *nature morte*—and the representation of the panic about virility and generation in Hemingway's novel *The Sun Also Rises* (1926). I have in mind in particular the scene in that novel which reads that panic through the display of stuffed dogs in a taxidermist's shop window, a scene that concisely draws into relation the still life of the market and the skin games of naturalism: "Simple exchange of values. You give them money. They give you a stuffed dog."[44]

Jack London's *The Sea-Wolf* tells the story of the regeneration of the "effeminate" writer, Humphrey Van Weyden, through his trials aboard the ship captained by the eponymous Wolf Larsen. The work of the ship is the taking of skins. Following the migration of the rookeries or breeding colonies, the ship "travelled with it, ravaging and destroying, flinging the naked carcasses to the shark and salting down the skins so that they might later adorn the fair shoulders of the women of the cities" (*SW* 603). The making of Van Weyden into a man ("a master of matter") involves not merely the slaughter of breeding seals ("a flank attack on the nearest harem" [*SW* 709]) but, above all, the solution to a problem in mechanics—the resetting of the fallen masts of the ship.

> But where were we to begin? If there had been one mast standing, something high up to which to fasten blocks and tackles! But there was nothing. It reminded me of the problem of lifting oneself by one's bootstraps. I understood the mechanics of levers; but where was I to get a fulcrum? (*SW* 732)

Not surprisingly, the Archimedean paradox is also the paradox of the *self-made man*. The problem in mechanics is the appropriate form of the making of men in machine culture, the form of virgin birth proper to that culture. As Van Weyden's "mate" Maud declares: "I can scarcely bring myself to realize that that great mast is really up and in; that you have lifted it from the water, swung it through the air, and deposited it here where it belongs" (*SW* 758). This is what sex in machine culture looks like.

But the making of Van Weyden parallels what seems to be the *un*making of Wolf Larsen. Larsen is at once sadistic (his command was "like the lash of a whip" [*SW* 497]) and the perfect specimen of the natural body of man ("I had never before seen him stripped, and the sight of his body quite took my breath away" [*SW* 593]). Larsen's power, however, is "a thing apart from his physical semblance" (*SW* 494). Stricken blind and progressively motionless in a process of degeneration that parallels Van Weyden's regeneration, Larsen is at once "imprisoned somewhere within that flesh" and in the process of "rising above the flesh" (*SW* 754). London's rudimentary lessons in the mind/body problem thus make visible the reduction of life to motion and to the natural body, *and* the transcendence of the natural that "proved our mastery over matter" (*SW* 767). If the flesh was quite dead, yet "the man of him was not changed" (*SW* 754): "it was disembodied. . . . It knew no body" (*SW* 764).

This process of transcendence is a process of transcending the natural and the female both. It is, finally, a matter of mechanics. "His body," London writes of Larsen, "thanks to his Scandinavian stock, was fair as the fairest woman's" (*SW* 593). But this fair and womanly skin is a natural cover that barely conceals something else entirely: "I remember his putting his hand up . . . and my watching the biceps move like a living thing under its white sheath" (*SW* 593). The power that moves "like" a living thing, under the white male skin, is thus only the semblance or reproduction of life. This male reproduction of life is thus revealed with the stripping off of that semblance of natural life and natural motion and with the achievement of unnatural life and unnatural motion: "I did not argue. I had seen the mechanism of the primitive fighting beast, and I was as strongly impressed as if I had seen the engines of a great battleship or Atlantic liner" (*SW* 594). This is what the unnatural body of man looks like. The stripping away of skins is thus the perpetual reminder of the difference between the natural and the technological. But it is ultimately and more powerfully the rewriting of this difference: the perpetual redemonstration of the unnaturalness of Nature—of "the mechanism of the primitive." "Under the skin," Artaud writes, "the body is an over-heated factory."[45]

Notes

Introduction: Case Studies and Cultural Logistics

1. It scarcely needs noting that the rich spreading of the topics of the body and technology in recent cultural studies has continued to inform and stimulate this study since the first part, "The Naturalist Machine," was presented, in 1984, until the final part, "The Love-Master," was presented, in 1989; the route from first to last has proved more direct in retrospect than in transit. Beyond the many local debts accrued and noted in what follows, these widely interdisciplinary chapters are more generally and indirectly indebted to the range of stimulating work that has emerged alongside, at times in response to, and at times along other tracks than this study takes and, above all, to the emerging idioms of understanding such work has made possible.
2. Bruno Latour, *The Pasteurization of France*, trans. Alan Sheridan and John Law (Cambridge, MA: Harvard University Press, 1988), p. 183. The principle of irreduction, insistent for instance throughout "Foucauldian" work on discursive practices, is remarkably absent from a good deal of cultural analyses; such analyses, that is, are remarkably inattentive to the notion that things are, among other things, what they appear to be.
3. Robert Lincoln O'Brien, "Machinery and English Style," *Atlantic Monthly* 94 (1904), 464.
4. Samuel Langhorne Clemens (Mark Twain), *A Connecticut Yankee in King Arthur's Court*, ed. Allison R. Ensor (New York: Norton, 1982); subsequent references are to this Norton Critical Edition and are included parenthetically in the text.

5. I am relying on here: James M. Cox, "*A Connecticut Yankee in King Arthur's Court*: The Machinery of Self-Preservation," incl. in the Norton Critical Edition, pp. 390–401; John F. Kasson, *Civilizing the Machine: Technology and Republican Values in America, 1776–1900* (Harmondsworth: Penguin, 1976), pp. 202–15; Justin Kaplan, *Mr. Clemens and Mark Twain* (New York: Simon and Schuster, 1966); Tom Burnam, "Mark Twain and the Paige Typesetter: A Background for Despair," *Western Humanities Review* 6 (1951), 29–36. On American machine culture more generally, in addition to sources elsewhere cited, see Cecilia Tichi's richly informed *Shifting Gears: Technology, Literature, Culture in Modernist America* (Chapel Hill: University of North Carolina Press, 1987); Thomas P. Hughes, *American Genesis: A Century of Invention and Technological Enthusiasm 1870–1970* (New York: Penguin, 1989); and, among earlier standard accounts: Lewis Mumford, *Technics and Civilization* (New York: Harcourt, Brace, 1934); Siegfried Giedion, *Mechanization Takes Command* (1948; repr. New York: Norton, 1969); Leo Marx, *The Machine in the Garden* (New York: Oxford University Press, 1964).
6. Twain, letter to Theodore Crane, October 5, 1888; as quoted by James M. Cox, "The Machinery of Self-Preservation," p. 397.
7. Beyerlen, as quoted by Friedrich A. Kittler, *Discourse Networks 1800/1900*, trans. Michael Metteer, with Chris Cullens (Stanford: Stanford University Press, 1990), p. 195.
8. Kittler, *Discourse Networks 1800/1900*, pp. 195, 212. The linking of hand, eye, and letter in the act of writing by hand posits the circular translation from mind to hand to eye and hence from the inward and invisible to the outward and visible and physical. This projects what Kittler describes as the discourse network of 1800: the notion of "the continuous transition from nature to culture" and thus a continuous transition from the self that writes and the writing of the self. By this logic, the work of writing and the cultural work of self-making refer back to (translate) each other at every point. A version of this notion of self-writing is conserved also in some recent work on American realist writing, most significantly and most provocatively, the work of Michael Fried and Walter Benn Michaels. (See Fried, *Realism, Writing, Disfiguration: On Thomas Eakins and Stephen Crane* [Chicago: University of Chicago Press, 1987] Michaels, *The Gold Standard and the Logic of Naturalism* [Berkeley: University of California Press, 1987].) Together Fried and Michaels provide fascinating instances of scenes of writing in turn-of-the-century American realist texts. But proceeding largely without reference to the new technologies of writing at the turn of the century and their effects, and, in effect, presupposing the circuit of hand-letter-self in

the act of writing by hand, these accounts reconstruct a tautological relation between writing and identity: a circular relation such that the writer's self-absorption refers to the scene of writing which refers to the writer's self-absorption. Hence this reconstructive account can only acknowledge the naturalist deconstruction or disarticulation of the links between writing and identity in terms of a crisis of self-making and self-possession: that is, in terms of the vicissitudes of possessive individualism. (The understanding of liberal market relations as normative is explicit in Michaels's account, as is the equation between writing and "owning" the self.) These accounts participate in, I would suggest, what Mark Poster has traced as the general move to reincorporate new writing-technologies and the mode of "information within the market system." (*The Mode of Information: Poststructuralism and Social Context* [Chicago: University of Chicago Press, 1990], p. 159.) That is, they instance the attempt to rewrite what Kittler specifies as the naturalist discourse network of 1900 in the anterior and anachronistic terms of what Kittler specifies as the romantic discourse network of 1800. More precisely, they instance the move to rewrite what I will be describing here as machine culture and disciplinary individualism in the terms of market culture and possessive individualism. In what follows I will be concerned with the recalcitrant appeals of such "reconstructions" of market culture (with "the romance of the market in machine culture": see Part II). And I will centrally be concerned with some of the ways in which new writing-technologies, processes of representation, and transformations in control-technologies and work processes operate in the discourses and practices of machine culture.

9. Kittler, *Discourse Networks*, p. 224.

10. William Dean Howells, *Harpers Weekly*, Vol. XXXII (January 14, 1888), p. 23. See also Thomas P. Lockwood, "Electrical Killing," *The Electrical Engineer*, Vol. VII (March 1888), p. 89. On the question of the telephone technology, see Avital Ronell's inventive and provocative *The Telephone Book: Technology, Schizophrenia, Electric Speech* (Lincoln: University of Nebraska Press, 1989). On the understanding of the electric current ("the deadly current") in relation to work and war technologies, see Jeffrey Herf, *Reactionary Modernism: Technology, Culture, and Politics in Weimar and the Third Reich* (Cambridge: Cambridge University Press, 1984). For a more comprehensive account of the relays between electric signals and signs and the body-machine complex, see my "Edison's Body Electric" (forthcoming).

11. Michel Chevalier, *Society, Manners, and Politics in the United States: Letters on North America* (Ithaca: Cornell University Press, 1961), p. 270.

12. On Taylorism and on the scouting movement's making of men, see Part V.

13. On the "coupling" of bodies and machines, see Part I. As Dolf Sternberger puts it, in his excellent *Panorama of the Nineteenth Century*, trans. Joachim Neugroschel (1955; repr. New York: Urizen, 1977): "the steam engine appears as a 'copulation' or union of a natural and an artificial element, whereby the disturbing factor of relative unpredictability, which the statistician must constantly try to restrain, is attributed to nature," p. 19. On the natural evolution and reproduction of machines, see Samuel Butler, *Erewhon* (1872; repr. Harmondsworth: Penguin, 1935), particularly chaps. 23–25, "The Book of the Machines," and see also the remarkably similar account set out in Hans Morajec's recent *Mind Children: The Future of Robot and Human Intelligence* (Cambridge, MA: Harvard University Press, 1988).

14. Jack London, "The Apostate," included in the Library of America edition of *Jack London: Novels and Stories* (New York, 1982), p. 801. ("The Apostate" was first published in *Women's Home Companion*, September 1906.)

15. See Anson Rabinbach, *The Human Motor: Energy, Fatigue, and the Origins of Modernity* (New York: Basic Books, 1990), pp. 19–42, 146–78.

16. Rabinbach, *The Human Motor*, p. 167.

17. On the body as energy-converting, thermodynamic machine and on the interconversion of living bodies and machines, see, for instance, Hermann von Helmholtz's highly influential, "On the Interaction of Natural Forces" (1854), reprinted in *Popular Lectures on Scientific Subjects*, trans. E. Atkinson (London: Longmans, Green, 1908) and his "The Conservation of Force: A Physical Memoir" (1847), in *Selected Writings of Hermann von Helmholtz*, ed. Russell Kahn (Middletown, CT: Wesleyan University Press, 1971). See also Dolf Sternberger, "Natural/Artificial," in his *Panorama of the Nineteenth Century*, pp. 29–38; Anson Rabinbach, *The Human Motor*, pp. 56–68.

18. Cited in Wolfgang Schivelbusch, *The Railway Journey: The Industrialization of Time and Space in the Nineteenth Century* (Berkeley: University of California Press, 1986), p. 124.

19. On the American habit of calculation, see Patricia Cline Cohen, *A Calculating People: The Spread of Numeracy in Early America* (Chicago: University of Chicago Press, 1982). On the calibrations of the natural body at the turn of the century, see in addition to the sources already cited: Stephen Jay Gould, *The Mismeasure of Man* (New York: Norton, 1981); Daniel J. Kevles, *In the Name of Eugenics: Genetics and the Uses*

of Human Heredity (Berkeley: University of California Press, 1985); Sharon E. Kingsland, *Modeling Nature: Episodes in the History of Population Ecology* (Chicago: University of Chicago Press, 1985). On the links between the calculus of probabilities, statistics, and psychophysics, see Stephen M. Stigler, *The History of Statistics* (Cambridge, MA: Harvard University Press, 1986), pp. 239–61.

20. On the transformations in the relations between the machine process and information-processing, see James R. Beniger, *The Control Revolution: Technological and Economic Origins of the Information Society* (Cambridge, MA: Harvard University Press, 1986).
21. Jack London, *John Barleycorn,* Library of America edition of *Jack London: Novels and Social Writings* (New York, 1982), p. 967.
22. Etienne-Jules Marey, *La Méthode graphique dans les sciences expérimentales et principalement en physiologie et en médicine* (Paris: G. Masson, 1878), p. iv.
23. Rabinbach, *The Human Motor,* p. 97.
24. Marey, *La Méthode graphique*, p. 108.
25. Ronald E. Martin, *American Literature and the Universe of Force* (Durham, NC: Duke University Press, 1981), p. 186.
26. See, for instance, Leo Marx, *The Machine in the Garden*, the point of which is in part that whenever a man and a woman are alone in a nineteenth-century American novel, a train goes by. On the erotics of steam technology in the naturalist novel, see the following chapter.
27. Not the least striking is the series of photographs, included in the 1891 edition of Muybridge's work, of a naked female model in whom hysterical convulsions have been induced/simulated. On the connections at work here between hysteria and the visual pleasure of looking and recording, see Linda Williams, *Hard Core: Power, Pleasure, and the "Frenzy of the Visible"* (Berkeley and Los Angeles: University of California Press, 1989), pp. 37–53; on motion and hysteria, see Monique David-Menard, *Hysteria from Freud to Lacan: Body and Language in Psychoanalysis,* trans. Catherine Porter (Ithaca: Cornell University Press, 1989), esp. chap. 4 "Is a Metapsychology of Movement Possible?"
28. Michel de Certeau, "Railway Navigation and Incarceration," in *The Practice of Everyday Life*, trans. Steven Randall (Berkeley and Los Angeles: University of California Press, 1984), p. 113.
29. On the phenomenology of railway transport in the nineteenth century, see Michel de Certeau, "Railway Navigation and Incarceration"; Dolf Sternberger, *Panorama of the Nineteenth Century*; Wolfgang Schivelbusch, *The Railway Journey: The Industrialization of Time and Space in the Nineteenth Century.*

30. Emile Zola, *La Bête humaine*, trans. Leonard Tancock (Harmondsworth: Penguin, 1977), p. 30. On human-machines systems, see Gilles Deleuze and Félix Guattari, *A Thousand Plateaus: Capitalism and Schizophrenia*, trans. Brian Massumi (Minneapolis: University of Minnesota Press, 1987), pp. 453–59 et passim.
31. Emile Zola, *La Bête humaine, Oeuvres completes* (Paris: Fasquelle, 1967), p. 43.
32. As Slavoj Žižek somewhat differently puts it: "In the Lacanian perspective, we should change the terms and designate as the most 'cunning' ideological procedure the very opposite eternalization: *overly-rapid historicization* . . . In other words, if over-rapid universalization produces a quasi-universal Image whose function is to make us blind to its historical, socio-symbolic determination, over-rapid historicization makes us blind to the real kernel which returns as the same through diverse historicizations/symbolizations." *The Sublime Object of Ideology* (London and New York: Verso, 1989), p. 50.
33. Latour, *The Pasteurization of France*, p. 167.

Part I: The Naturalist Machine

1. Frank Norris, *The Octopus: A Story of California* (1901), repr. in *The Complete Edition of Frank Norris*, 10 vols. (Garden City, NY: Doubleday, Doran, & Co., 1928), 2:21. All subsequent references to this work are to this edition, vols. 1 and 2, and are given parenthetically by volume and page number in the text.
2. This move to the Asian marketplace is managed by the industrialist-turned-shipper Cedarquist, who earlier notes that if the "great word" of the nineteenth century was production, "the great word of the twentieth century will be . . . markets" (2:21). The turn to Asia as the new marketplace for American overproduction displays not merely a global and imperialist marketing strategy but also a promotion and exploitation of late nineteenth-century anxieties about masculinity. As the social historian Ronald T. Takaki argues in *Iron Cages: Race and Culture in Nineteenth-Century America* (New York: Alfred A. Knopf, 1979), the imperial outlet for American surplus explicitly was seen as a violent reassertion of the racial and social "body," as a "masculine thrust toward Asia" (pp. 253–79).
3. On the agrarianist antimony of producer and speculator, and the economic and aesthetic problems it entails, see Walter Benn Michael's excellent "Dreiser's *Financier*: The Man of Business as a Man of Letters," in

American Realism: New Essays, ed. Eric J. Sundquist (Baltimore and London: The Johns Hopkins University Press, 1982), pp. 278–95.

4. Perry Miller, *Life of the Mind in America: From the Revolution to the Civil War* (New York: Harcourt, Brace & World, 1965), p. 293; J.A. Meigs, as quoted by Miller, "The Responsibility of Mind in a Civilization of Machines," *American Scholar* 31 (Winter 1961), 62.
5. Takaki, *Iron Cages*, p. 149.
6. Augustus K. Gardner, *History of the Art of Midwifery* (New York: Stringer & Townshend, 1852), p. 1.
7. G.J. Barker-Benfield, *The Horrors of the Half-Known Life: Male Attitudes Toward Women and Sexuality in Nineteenth-Century America* (New York: Harper & Row Publishers, 1976), p. 265.
8. "Animalcule" is a term for microscopic animals but also, and especially before the spermatic function in reproduction was determined (that is, before Oskar Hertwig's account of fertilization in 1876), for spermatozoa. For a brief summary of late nineteenth-century theories of reproduction and fertilization, see Patrick Geddes and J. Arthur Thomson, *The Evolution of Sex* (1889; repr., London and New York: Scribner's 1899), pp. 156–67.
9. For an informed inventory of naturalist notions of force, see Ronald E. Martin, *American Literature and the Universe of Force* (Durham, NC: Duke University Press, 1981); see also William H. Jordy, *Henry Adams: Scientific Historian* (New Haven and London: Yale University Press, 1952). Harold Kaplan's recent *Power and Order: Henry Adams and the Naturalist Tradition in American Fiction* (Chicago and London: University of Chicago Press, 1981) tends, like most treatments of American naturalism, to relay rather than examine naturalist "ideologies" of force, taking as its premise a radical antimony between literary and social forms of power, between art and power, and automatically assuming, for instance, that "even the most sophisticated political theory works at odds with the literary imagination" (p. ix). Important discussions of Norris include: Donald Pizer, *The Novels of Frank Norris* (Bloomington: Indiana University Press, 1966); and the chapters on Norris in Maxwell Geismar, *Rebels and Ancestors: The American Novel, 1890–1915* (Boston: Houghton Mifflin Co., 1953); Kenneth Lynn, *The Dream of Success: A Study of the Modern American Imagination* (Boston: Little, Brown & Co., 1955); and Larzer Ziff, *The American 1890s: Life and Times of a Lost Generation* (New York: Viking Press, 1966).
10. On the centrality of the thermodynamic model in the nineteenth century, with particular reference to Zola, see Michel Serres's extraordinary *Feux et signaux de brume, Zola* (Paris: Grasset, 1975), and also his *Hermes:*

Literature, Science, Philosophy, ed. Josué V. Harari and David Bell (Baltimore and London: The Johns Hopkins University Press, 1982).

11. Henry Adams, "A Letter to American Teachers of History," in *The Degradation of the Democratic Dogma* (New York: Macmillan Co., 1920), pp. 144, 140.

12. Thorstein Veblen, *The Theory of the Leisure Class: An Economic Study of Institutions* (1899; repr., New York: New American Library, 1953), pp. 27–28. The difficulty in securely locating the category of *production* in machine culture reappears throughout Veblen's writings on economics: for instance, in his essays on "The Preconceptions of Economic Science." In his account of the physiocrat conception of production, for example, Veblen quotes J.-B. Say's "last resort" notion of the nature of production: "Dieu seul est producteur. Les hommes travaillent, receuillent, économisent, conservent; mais *économiser* n'est pas *produire*." (See Thorstein Veblen, *The Place of Science in Modern Civilization and Other Essays* [New York: B.W. Huebsch, 1919], pp. 92, 126.) By the physiocrat account God only produces and the only ground of economic reality is "the nutritive process of Nature." By Veblen's account, the difference between production and conversion is the difference between "the effort that goes to create a new thing with a new purpose given to it by the fashioning hand of its maker out of passive ('brute') material; while exploit, so far as it results in an outcome useful to the agent, is the conversion to his own ends of energies previously directed to some other end by another agent" (*Theory of the Leisure Class*, pp. 27–28). But this residually "creationist" account of production, and its personifying and animistic effects ("the fashioning hand of its maker"), is directly at odds with Veblen's (equally insecure) account of "agency" and its relation to "the machine process." As Veblen puts it, immediately following this distinction between "industry" and "exploit": "As a matter of selective necessity, man is an agent. He is, in his own apprehension, a center of unfolding impulsive activity—'teleological' activity. He is an agent seeking in every act the accomplishment of some concrete, objective, impersonal end. By force of his being such an agent he is possessed of a taste for effective work, and a distaste for futile effort" (p. 29). It is not entirely clear here whether agency is anything more than the "apprehension" or self-projection of agency—not least because the effect of agency is an effect of "the necessity" the agent is "by force . . . possessed of," not of the force he possesses. Beyond that, persons appear as agents in the act of acting "impersonally" and in their "distaste of futile effort": for instance, the futile effort involved in projecting notions of the creator's "strong hand"—the "naive, archaic habit of construing all manifestations of force in terms of personality or 'will power' " (p. 31). The naive and

archaic habit of construing force in terms of personality is thus the naive and archaic habit of personifying agencies or forces acting impersonally: that is, the habit of personifying persons. What acts impersonally and produces is, by Veblen's account, thus epitomized by the machine process. And if persons are agents by reason of acting impersonally, what acts most like a person is, for Veblen, the machine process. (On some further implications of Veblen's renegotiations of the shifting line between persons and the machine process, see Parts II and V.)

13. Veblen, *Theory of the Leisure Class*, p. 28. Women function as something of a shifter term and even "fall guy" in Veblen's evolutionary theory, representing at once productive industry and the incitement to exploit and waste.
14. Henry Adams, *The Education of Henry Adams* (1918; repr., Boston: Houghton Mifflin Co., 1946), pp. 380, 453, 455–56, 344.
15. Adams, "A Letter to American Teachers of History," p. 231.
16. Adams, *The Education of Henry Adams*, p. 387.
17. Theodore Dreiser, *The Financier* (New York: Thomas Y. Crowell Co., 1974), p. 61.
18. I refer here to the notion of "miraculated production" articulated by Gilles Deleuze and Félix Guattari in their discussions of production-machines and of the body as machine in *Anti-Oedipus: Capitalism and Schizophrenia*, trans. Robert Hurley, Mark Seem, and Helen R. Lane (Minneapolis: University of Minnesota Press, 1983), esp. chap. 1.
19. Norris's notes and manuscripts are in the Bancroft Library, University of California, Berkeley.
20. Hilma's pregnancy is also juxtaposed, in this scene, to the slaughter of the hyperproductive rabbits; this massacre, in turn, reinvokes the earlier "massacre of innocents" (1:47)—the locomotive's slaughter of the sheep (1:47–48); Annixter, Norris writes, felt "an inane sheepishness when [Hilma] was about" (1:76–77).
21. On the significance of the Medusa figure, see Neil Hertz, "Medusa's Head: Male Hysteria under Political Pressure," *Representations* 4 (Fall 1983), 27–54.
22. Frank Norris, *McTeague*, ed. Donald Pizer (1899; repr. of first ed., New York: W.W. Norton & Co., 1977), p. 213. All subsequent references to this work are to this edition and are cited parenthetically by page number in the text.
23. Rev. John Todd, *The Sunset Land; or, The Great Pacific Slope* (Boston: Lee & Shepard, 1870), pp. 67, 47, 124. Barker-Benfield observes, in *Horrors of the Half-Known Life*, that Todd, among others, viewed "male activities of all kinds, from his own to those of California gold-miners,

as the emulation of woman's powers of gestation and parturition" (p. 217). In *The Death of Nature: Women, Ecology and the Scientific Revolution* (New York: Harper & Row, 1980), Carolyn Merchant traces the transformation of "the normative constraints against the mining of mother earth" into the normative sanctioning of mining: "Sanctioning mining sanctioned the rape or commercial exploration of the earth . . . The organic framework, in which the Mother Earth image was a moral restraint against mining, was literally undermined by the new commercial activity" (pp. 29–41).

24. Karl Marx, *Capital* (New York: International Publishers, 1967), 3:827.
25. Frank Norris, *Vandover and the Brute* (1914; repr. of first ed., Lincoln: University of Nebraska Press, 1978), p. 10. All subsequent references to this novel are to this edition and are cited parenthetically by page number in the text.
26. Vandover's art is consistently allied to mechanics; the sites of art in the novel, for instance, are preeminently the Mechanics' Library and the Mechanics' Fair; and Vandover's final artistic work involves the painting of little pictures—of steamships, for example—on the lacquered surfaces of iron safes (p. 314).
27. Joseph Le Conte, "Correlation of Vital with Chemical and Physical Forces," in Balfour Stewart et al., *The Conservation of Energy* (New York: D. Appleton, 1875), pp. 179, 190. Le Conte's naturalist linking of generation and degradation is not at all atypical. See, for instance: Henry Fiske, "The Dependence of Life on Decomposition" (1871), included with the *Pamphlets on Evolution* in the Le Conte collection, Biology Library, University of California, Berkeley. As François Jacob observes in the course of his invaluable history of biology, *The Logic of Life*, "The concepts of thermodynamics completely upset the notion of a rigid separation between beings and things, between the chemistry of the living and laboratory chemistry. With the concept of energy and that of conservation, which united the different forms of work, all the activities of an organism could be derived from its metabolism. . . .the same elements compose living beings and inanimate matter; the conservation of energy applies equally to events in the living and in the inanimate world." (*The Logic of Life: A History of Heredity*, trans. Betty E. Spillman [New York: Pantheon Books, 1973], p. 194.) See also: David F. Channell, *The Vital Machine: A Study of Technology and Organic Life* (New York: Oxford University Press, 1991).
28. Le Conte, "Correlation of Vital with Chemical and Physical Forces," p. 192.
29. Geddes and Thomson, *Evolution of Sex*, pp. 26, 234. On Geddes, see Jill Conway, "Stereotypes of Femininity in a Theory of Sexual Evolution,"

in *Suffer and Be Still: Women in the Victorian Age*, ed. Martha Vicinus (Bloomington and London: Indiana University Press, 1972), pp. 140–54; Philip Boardman, *Patrick Geddes: Maker of the Future* (Chapel Hill: University of North Carolina Press, 1944); and Boardman's *The Worlds of Patrick Geddes* (London: Routledge & Kegan Paul, 1978). It should be noted that the connection between heat and sexual differentiation and reproduction goes back at least to Galen and Aristotle. (See, for instance: Galen, *On the Usefulness of the Parts of the Body*, ed. and trans. Margaret Tallmadge May [Ithaca: Cornell University Press, 1968].) My concern here is with the consequences of the shift from thermic to thermodynamic theories of sexual biology.

30. See my *Henry James and the Art of Power* (Ithaca: Cornell University Press, 1984), esp. the "Postscript," pp. 171–95.

31. Jacques Donzelot, *The Policing of Families*, trans. Robert Hurley (New York: Pantheon Books, 1979), p. 232.

32. Donzelot, *Policing* , p. 6. See also, Michel Foucault, "The Subject and Power," *Critical Inquiry*, 8 (Summer 1982). The investment in a radical opposition between sexual and economic domains figures prominently even in, or especially in, attempts to locate points of intersection between them—for instance, in attempts to find the missing link between Marxist and feminist analyses. Thus, Catherine A. MacKinnon, in her richly suggestive "Feminism, Marxism, Method, and the State: An Agenda for Theory," in *The Signs Reader: Women, Gender, and Scholarship*, ed. Elizabeth Abel and Emily K. Abel (Chicago and London: University of Chicago Press, 1983), pp. 227–56, begins by restating such an opposition: "Sexuality is to feminism what work is to marxism: that which is most one's own, yet most taken away." Although this antinomy is insisted upon throughout, MacKinnon further suggests that "instead of engaging the debate over which came (or comes) first, sex or class, the task for theory is to explore the conflicts and connections between the methods that found it meaningful to analyze social conditions in terms of those categories in the first place" (p. 239). I am suggesting that one form such an exploration of conflicts and connections might take is the biopolitical analysis I have outlined here. More locally, MacKinnon is certainly right in stating that a "synthesis" of sexual and economic categories cannot operate simply through an analogizing by which "the marxist meaning of reproduction, the iteration of productive relations, is punned into an analysis of biological reproduction" (p. 238). Rather, what I have tried to focus on here are the naturalist rewritings of production, the ways in which the contradictions and differences between forms of mechanical and biological reproduction become operational.

33. Geddes and Thomson, *Evolution of Sex*, p. 286.

34. Josiah Strong, *The Times and Young Men* (New York: Baker and Taylor, 1901), p. 125.
35. Josiah Strong, *Our Country* (1891; repr., Cambridge, MA: Harvard University Press, 1963), p. 164.
36. See, for instance: T.J. Jackson Lears, *No Place of Grace: Antimodernism and the Transformation of American Culture, 1880–1920* (New York: Pantheon Books, 1981), and Lears's "From Salvation to Self-Realization: Advertising and the Therapeutic Roots of the Consumer Culture, 1880–1930," in *The Culture of Consumption: Critical Essays in American History, 1880–1980*, ed. Richard Wightman Fox and Lears (New York: Pantheon Books, 1983), pp. 3–38.
37. Michel Foucault, *Discipline and Punish: The Birth of the Prison,* trans. Alan Sheridan (New York: Pantheon Books, 1977), pp. 160–61.
38. For a more detailed treatment of this realist narrative supervision, see my "*The Princess Casamassima*: Realism and the Fantasy of Surveillance," *Nineteenth-Century Fiction* 35, 4 (March 1981), 506–34; repr. in *Henry James and the Art of Power.*
39. Frank Norris, "The Mechanics of Fiction," in *Blix: Moran of the Lady Letty: Essays on Authorship* (New York: P.F. Collier & Son Publishers, 1899), pp. 314, 316; repr. in Norris, *The Responsibilities of the Novelist, Complete Edition* 7:114, 116–17.

Part II: Physical Capital: The Romance of the Market in Machine Culture

1. George Fitzhugh, *Cannibals All! or Slaves Without Masters*, ed. C. Vann Woodward (Cambridge, MA: Harvard University Press, 1960), p. 258. Harriet Beecher Stowe, *Uncle Tom's Cabin* (Harmondsworth: Penguin, 1981), pp. 480–81. Subsequent references are to this edition and are cited parenthetically by page number in the text.
2. Charlotte Perkins Gilman, "The Yellow Wallpaper," in *Charlotte Perkins Gilman Reader* (New York: Pantheon, 1980), pp. 7–8.
3. Henry James, *The American*, ed. James W. Tuttleton (New York: Norton, 1978), p. 303. Subsequent references to the novel and preface are to this edition, which follows the London Macmillan edition of 1879, and are cited parenthetically in the text.
4. Standard accounts of the culture of consumption, and in terms of such a narrative of the fall (from things to representations, from substance to shadow or image, from production to consumption, from use to

exchange—from, most crudely, nature to culture and from bodies to artifacts, etc.) include, with important variants, of course: Daniel Bell, *The Cultural Contradictions of Capitalism* (New York: Basic Books, 1976); Christopher Lasch, *The Culture of Narcissism: American Life in An Age of Diminishing Expectations* (New York: Norton, 1979); Stuart and Elizabeth Ewen, *Channels of Desire: Mass Images and the Shaping of American Consciousness* (New York: McGraw Hill, 1982); Stuart Ewen, *All Consuming Images: The Politics of Style in Contemporary Culture* (New York: Basic Books, 1988); T.J. Jackson Lears, *No Place of Grace: Antimodernism and the Transformation of American Culture, 1880–1920* (New York: Pantheon Books, 1981).

5. Adam Smith, *An Inquiry into the Nature and Causes of the Wealth of Nations* (London: Methuen, 1904), 1:12. For a provocative account of the relations between philosophical and economic speculation, and the divisions between forms of doing and forms of looking, see Alfred Sohn-Rethel, *Intellectual and Manual Labor: A Critique of Epistemology* (Atlantic Highlands, NJ: Humanities Press, 1978).
6. On the "replacement" of the market system by the control-technologies of the machine process, see James R. Beniger, *The Control Revolution: Technological and Economic Origins of the Information Society* (Cambridge, MA: Harvard University Press, 1986). Beniger's account supplements and extends Alfred D. Chandler, Jr., *The Visible Hand: The Managerial Revolution in American Business* (Cambridge, MA: Harvard University Press, 1977).
7. I refer here to Pierre Bourdieu's notion of "cultural capital," as set out in *Distinction: A Social Critique of the Judgement of Taste* (Cambridge, MA: Harvard University Press, 1984).
8. On the representation of hotel-civilization in James's *The American Scene*, see my *Henry James and the Art of Power* (Ithaca: Cornell University Press, 1984), pp. 96–145. Just as the hotel or railway "hotel-car" in what James will later call American "hotel-civilization" makes possible a mobile and public market in private spaces (home away from home), the glass wall of the hotel lobby make visible the spectacle of consumption (a viewing screen that is also, as it were, an early version of "home shopping"). On plate glass and glass walls, see Wolfgang Schivelbusch, *The Railway Journey: The Industrialization of Time and Space in the Nineteenth Century* (Berkeley: University of California Press, 1986), pp. 45–51.
9. In *The Bourgeois and the Bibelot* (New Brunswick: Rutgers University Press, 1984), Remy G. Saisselin discusses the "bibelotization" of both art and women in the bourgeois consumerist aesthetic. He points as well to the close "parallels between museums and department stores" in later

nineteenth-century consumer society (pp. 33–49). Newman's viewing of the stream of girls through the huge wall of plate glass clearly invokes the department store technology of making commodities visible, the display of objects through the recent innovation of the glass wall. And Claire is, of course, both the best article in the market and an *objet d'art* for Newman. Saisselin further notes that "the flâneur might escape the lure of consumer goods merely by stepping from the Magasin du Louvre into the Musée du Louvre, to stroll, gaze, and lounge. Yet even here he might be attracted by objects—objects beyond his desire only because they could not be purchased" (p. 41). Newman's consumption of copies goes this distinction one better: "I have just bought a picture," he tells Tristram. " 'Bought a picture?' said Mr. Tristram, looking vaguely round at the walls [of the Musée du Louvre]. 'Why, do they sell them?' " (p. 27). I will discuss the scene of the museum, copying, and buying in a moment. My concern, however, is less with the world as department store (for example, Zola's *Au Bonheur des dames* [1883]) or with the representation of desire as the desire for representation in consumer culture (for example, Dreiser's *Sister Carrie* [1900]) than in the assumptions about the artifactuality of persons and historicity of bodies and desires that such accounts instance and that James's *The American* begins to imagine. In addition to Saisselin on consumerism, see Jean-Christophe Agnew's "The Consuming Vision of Henry James," in *The Culture of Consumption: Critical Essays in American History, 1880–1980*, ed. Richard Wightman Fox and T.J. Jackson Lears (New York: Pantheon, 1983); Michael B. Miller, *The Bon Marché: Bourgeois Culture and the Department Store, 1869–1920* (Princeton: Princeton University Press, 1981); Rachel Bowlby, *Just Looking: Consumer Culture in Dreiser, Gissing, and Zola* (New York: Methuen, 1985); Rosalind H. Williams, *Dream Worlds: Mass Consumption in Late Nineteenth-Century France* (Berkeley: University of California Press, 1982).

10. James's review, which appeared in *The Nation* 20 (May 6, 1875), 318–19, is reprinted in the Norton Critical Edition of *The American*, pp. 324–26. In *Notes on Paris*, Taine observes that "Women and works of art are related creatures . . . What is really wanted of them is possession or exhibition." *Notes sur Paris* (Paris: Hachette, 1901), p. 307. (My translation.)

11. Edward Bellamy, *Looking Backward, 2000–1887* (Cambridge, MA: Harvard University Press, 1967), p. 307.

12. David M. Potter, *People of Plenty: Economic Abundance and the American Character* (Chicago: University of Chicago Press, 1954), pp. 118–19. On the history and effects of American mass production and mechanical reproduction, see among other standard accounts of American making,

production, and standardization: David A. Hounshell, *From the American System to Mass Production 1800–1932: The Development of Manufacturing Technology in the United States* (Baltimore: The Johns Hopkins University Press, 1984); Daniel Boorstin, *The Americans: The Democratic Experience* (New York: Random House, 1973); *Yankee Enterprise: The Rise of the American System of Manufactures,* eds. Otto Mayr and Robert C. Post (Washington, D.C.: Smithsonian Institution Press, 1981). On imitation and reproducibility at the turn of the century, see Miles Orvell's recent *The Real Thing: Imitation and Authenticity in American Culture, 1880–1940* (Chapel Hill: University of North Carolina Press, 1989).

13. Richard Poirier provides an acute but ultimately alternative account of the Jamesian "type" in his finely perceptive treatment of *The American* in *The Comic Sense of Henry James: A Study of the Early Novels* (New York: Oxford University Press, 1960), pp. 44–94. See also, on typicality in James, William Veeder's impressively informed, *Henry James—The Lessons of the Master: Popular Fiction and Personal Style in the Nineteenth Century* (Chicago: University of Chicago Press, 1975), pp. 106–83.
14. See, for instance, Georg Simmel's essay on "Fashion" (1904), included in *On Individuality and Social Forms: Selected Writings,* ed. Donald N. Levine (Chicago: University of Chicago Press, 1971). The tension between "the special" and "the general" that for Simmel defines fashion is also the prototype for the tension between "individuality and social forms" that defines, for Simmel, the sociological *as such.*
15. The phrase in quotation marks is drawn from Thorstein Veblen's *Theory of the Leisure Class: An Economic Study of Institutions* (1899; repr., New York: New American Library, 1953), p. 55.
16. Veblen, *Theory of the Leisure Class*, pp. 63, 107.
17. On the American physical culture movement, see Harvey Green, *Fit for America: Health, Fitness, Sport, and American Society* (New York: Pantheon, 1986); T.J. Jackson Lears, "From Salvation to Self-Realization: Advertising and the Therapeutic Roots of the Consumer Culture, 1880–1930," in *The Culture of Consumption: Critical Essays in American History, 1880–1980*, eds. Richard Wightman Fox and T.J. Jackson Lears; John Higham, "The Reorientation of American Culture in the 1890s," in *Writing American History—Essays on Modern Scholarship* (Bloomington: Indiana University Press, 1970), pp. 73–102. I take up in some detail the relation between the natural and the national body in turn-of-the-century American naturalism in the last part of this study, "The Love-Master."
18. The most comprehensive account of the general shift in the notions of

"the standard" and "the normal" is George Canguilhem, *On the Normal and the Pathological*, trans. C. R. Fawcett (Dordrecht: D. Reidel, 1978); and the most comprehensive account of the effects of normalization and standardization on notions of individuality and systemic management, Michel Foucault, *Discipline and Punish: The Birth of the Prison*, trans. Alan Sheridan (New York: Pantheon Books, 1977). In *Henry James and the Art of Power*, I have attempted (in part by way of Canguilhem and Foucault) to set out some of the ways in which forms of normalization, standardization, and systemness are relayed by the form of the later nineteenth-century realist and naturalist novel, particularly as models of management and individualization in the disciplinary society. There I examine, for example, the statistical inscription of social spaces and individuals (chap. 1), the aesthetics of "the standard" and "the normal" (chap. 2), and the culture of systematic management (chap. 3). My concern here, in extending these accounts, is with the recalcitrant tension between market and systemic accounts of persons and its implications in consumer society.

19. On advertising as the "technology of romance" and as the "discipline of desire" at the turn of the century, see my *Henry James and the Art of Power*, pp. 142–45.

20. Igor Kopytoff, "The Cultural Biography of Things: Commodization as Process," in *The Social Life of Things: Commodities in Cultural Perspective*, ed. Arjun Appadurai (Cambridge: Cambridge University Press, 1986), p. 64.

21. See Lasch, *The Culture of Narcissism*, pp. 21, 119. et passim; Ewen and Ewen, *Channels of Desire* and Ewen, *All Consuming Images*. What the "normative relations between things and their representations" protect, for Lasch, is "the very idea of reality" (p. 160), that is, "the distinction between illusion and reality" (p. 159). This distinction is typically (and not merely in Lasch) mapped onto social and class distinctions. What scandalizes Lasch is not so much excess or abundance but its generalization or democratization. In short, "the masses" here represent something like a principle of realism or materiality or deep embodiment or "massiness," the reality principle against which the "culturalist" privilege of relative disembodiment, illusionism, or excess can be measured. Lasch's account "describes a way of life that is dying—the culture of competitive individualism" (p. 21) and that account everywhere makes visible how the market culture of competitive or possessive individualism allocates personhood in accord with a principle of scarcity and how "the idea of reality" and "representation" or "illusion" are drawn into relation to these invidious allocations of personhood, across class, race, and gender lines. For a more extended treatment of these problems, and particularly

of the entanglements of the aesthetics of realism and consumerism, see Part IV, "The Still Life."

22. Bell, *The Cultural Contradictions of Capitalism*, pp. v., 82, et passim. For Bell, as for Lasch, the crisis of consumer culture is the democratization of excess ("the cultural transformation of modern society is due, singularly, to the rise of mass consumption, or the diffusion of what were once considered luxuries to the middle and lower classes in society" [p. 65]); and what this entails is a crisis at once aesthetic and epistemological (the "erasure of the distinction between art and life" [p. xv]). In his usefully detailed, *The Morality of Spending: Attitudes toward the Consumer Society in America, 1875–1940* (Baltimore: The Johns Hopkins University Press, 1985), Daniel Horowitz fills in the dichotomies (morality/spending, ascetics/hedonism) that underwrite the narrative accounts provided by Lasch and Bell. Horowitz notes, particularly in his epilogue (pp. 166–71), the difficulties and hesitations in historicizing or periodizing consumption behavior and attitudes in these oppositional terms. I am more interested in examining some of the implications of those hesitations—and their relation to the uncertainties about the status of persons, bodies, and representations in consumer culture—than I am in rehearsing the stories that Bell, Lasch, Ewen, Lears, and Horowitz, among others, have been telling.
23. It is not hard to see, for example, that the notion of a fall from "use" to "exchange" over the course of the nineteenth century not merely ignores the fact that capitalist production *is* production for exchange but also, for instance, that turn-of-the-century marginal utilitarian economics centers on the category of "use." On some of these impasses in turn-of-the-century economics, see, for instance: Thorstein Veblen, "The Limitations of Marginal Utility," in *The Place of Science in Modern Civilization and Other Essays* (New York: B.W. Huebsch, 1919). For an incisive critique of the weak periodizations that continue to dominate accounts of the culture of consumption, see Catherine Gallagher, "Review Article," *Criticism* 29, 2 (Spring 1987), 233–42.
24. See Werner Sombart, *Luxury and Capitalism* (Ann Arbor: University of Michigan Press, 1967) and Arjun Appadurai's useful "Introduction: Commodities and the Politics of Value," in *The Social Life of Things*, pp. 3–63. For an excellent case history of the links between the body and economics (the bioeconomics) in consumption, see Sidney W. Mintz, *Sweetness and Power: The Place of Sugar in Modern History* (New York: Penguin, 1985).
25. Veblen, *The Theory of the Leisure Class*, pp. 33, 82.
26. For an extended account of the disciplines of the machine process, see Parts III ("Statistical Persons") and V ("The Love-Master").

27. Theodor W. Adorno, "Veblen's Attack on Culture," in *Prisms*, trans. Samuel and Sherry Weber (Cambridge, MA: MIT Press, 1981), pp. 75–94. On the contradictions in Veblen's account of consumption, see also Horowitz, *The Morality of Spending*, pp. 37–41.
28. Bourdieu, *Distinction*, pp. 5, xiii. Bourdieu everywhere aligns the position of the sociologist and the position of the working classes, who of necessity, for Bourdieu, express a realism in consumption, a "choice of the necessary" that enacts the systematic reduction Bourdieu's critique also works to perform. On the nexus of realism and consumerism, see Part IV, below. For a compelling reading of Bourdieu in relation to turn-of-the-century "ethnic" literature, see Phillip Barrish, " 'The Genuine Article': Cultural and Economic Capital in *The Rise of David Levinsky*" (forthcoming).
29. Bourdieu, *Distinction*, p. 1.
30. As Miles Orvell suggests in *The Real Thing: Imitation and Authenticity in American Culture, 1880–1940*, not merely is it the case that "the image was viewed as both a mimetic and an artificial construction, both specific and general" (p. 88). Beyond that, the individuality of the individual cannot be separated from the generic standards and classes, the double process of individualization and classification, made possible by the machine process. Conversely, as Orvell somewhat differently puts it, "material things had their qualities and classes, just as did people" (p. 44).
31. See W.F. Haug, *Critique of Commodity Aesthetics: Appearance, Sexuality, and Advertising in Capitalist Society*, trans. Robert Bock (Minneapolis: University of Minnesota Press, 1986): "The body, on whose behalf all this advertising is happening, adopts the compulsory traits of a brand-named product" (p. 83). On another version of the making of the American consumer, with particular emphasis on its racialized embodiments in a later period, see Lauren Berlant's superb "National Brands/National Body: *Imitation of Life*" in *Comparative American Identities: Race, Sex, and Nationality in the Modern Text*, ed. Hortense Spillers (New York: Routledge, 1990). For a third, more general, study of the bodily and the artifactual that informs both our accounts, albeit to different ends, see Elaine Scarry, *The Body in Pain: The Making and Unmaking of the World* (New York: Oxford University Press, 1985).
32. The literature on the separate spheres of domesticity and the market, and their reorientation, is of course extensive. For a recent reexamination of the separate spheres argument, see Mary P. Ryan, *Women in Public: Between Banners and Ballots, 1825–1880* (Baltimore: The Johns Hopkins University Press, 1990). My more local concern at this point, and particularly in the next two sections of this chapter, is with some of the

ways in which the reorientation of interior states (both domestic and psychological) toward the public sphere of consumption is recognized, managed, and embodied.

33. Sherry B. Ortner, "Is Female to Male as Nature is to Culture," in *Woman, Culture, and Society*, eds. Michelle Zimbalist Rosaldo and Louise Lamphere (Stanford: Stanford University Press, 1974), pp. 67–87. I am indebted here to Naomi Schor's stimulating study, *Reading in Detail: Aesthetics and the Feminine* (New York: Methuen, 1987). As Schor suggests, "the negative connotations of the feminine" cross categories on both sides of the nature/culture dichotomy, for example, the cross-categorizations visible in lists such as "the decorative, the natural, the impure, and the monstrous" (p. 45). These crossings of the nature/culture dichotomy and the sex/gender system are registered in what I will be calling the unnaturalness of nature in American naturalism.
34. Henry James, *Hawthorne* (1879; repr., Ithaca: Cornell University Press, 1966), pp. 99–105.
35. Nathaniel Hawthorne, *The House of the Seven Gables,* ed. Seymour L. Gross (New York: Norton, 1967), pp. 151–52. Subsequent references are to this edition and are cited parenthetically by page in the text.
36. Hawthorne here provides a sort of barnyard version of the "transcendental" correspondence between bodies or material things and the meanings they embody. On embodiment of Hawthorne, see Sharon Cameron's *The Corporeal Self: Allegories of the Body in Melville and Hawthorne* (Baltimore: The Johns Hopkins University Press, 1981), pp. 77–157. On heredity, biology, and the body in the later nineteenth-century novel, see Gillian Beer, *Darwin's Plots: Evolutionary Narrative in Darwin, George Eliot, and Nineteenth-Century Fiction* (London: Ark, 1983); Peter Morton, *The Vital Science: Biology and the Literary Imagination, 1860–1900* (London: Allen & Unwin, 1984).
37. Hawthorne's animistic domestic economy, like Stowe's, disavows the impersonality of things. "Phoebe, and the fire that boiled in the teakettle," for instance, "were equally bright, cheerful, and efficient, in their respective offices" (p. 76); Rachel Halliday "put a spirit into the food and drink she offered," and domestic objects "arrange" *themselves* as if "in obedience to a few gentle whispers" (*Uncle Tom's Cabin*, pp. 217–23). "The fact was, it was, after all, the THING that I hated" (*UTC*, p. 342), St. Clare says, and what St. Clare hates is not merely the fact of slavery but the fact of the impersonal THING itself. Crucially, the protest against turning persons into things (a critique of reification) involves also an insistence on turning things into persons (an imperative of personification). Such an emphasis is not, of course, limited to the domestic and sentimental novel or treatise. Oliver Wendell Holmes, for example, in

the first lecture of *The Common Law* (1880; repr., Boston: Little, Brown, 1963), focuses on the persistence in liability and property law of the "metaphysical confusion" that involves a "personification of inanimate nature" (pp. 30, 12).

38. On the sexualization of the nineteenth-century family, see Carroll Smith-Rosenberg, *Disorderly Conduct: Visions of Gender in Victorian America* (New York: Knopf, 1985) and the more general account of Michel Foucault, *The History of Sexuality*, vol. 1, trans. Robert Hurley (New York: Pantheon, 1978).

39. Albert O. Hirschman comments on the persistent noncommercial and specifically sexual senses of "commerce" in *The Passions and the Interests: Political Arguments for Capitalism before Its Triumph* (Princeton: Princeton University Press, 1977), pp. 61–63. Note also the terms of John Dewey's euphoric description of commerce in his "Pragmatic America" (1922): "Commerce itself, let us dare say it, is a noble thing. It is intercourse, exchange, communication, distribution, sharing of what is otherwise secluded and private." *Pragmatism and American Culture*, ed. Gail Kennedy (New York: Heath, 1950), p. 59.

40. The communication of stories is, for Hawthorne, inevitably the "open[ing of] a door that communicated with the shop" (p. 36) and the "solution" to the anxious relation between persons and things that Hawthorne devises is that of a commerce embodied in persons—the romance of the market itself. The differences between private and public, between the sexual and the economic, the family and the market: these "separate spheres" are opposed on part of their surface but communicate on another level. That is, the reciprocal autonomy protected by the notion of "separate sphere" is itself premised on the ideology of the market: the market competition of free and autonomous individuals. Habermas has concisely summarized what we might call the double discourse of separate spheres, and, more particularly, the manner in which the notion of autonomous individuals freely interacting *in* the market is rewritten as the autonomy and freedom of individuals *from* the market:

> Such an autonomy of private people, founded on the right to property and in a sense also realized in the participation in a market economy, had to be capable of being portrayed as such. To the autonomy of property owners in the market corresponded a self-presentation of human beings in the family. The latter's intimacy, apparently set free from the constraints of society, was the seal on the truth of a private autonomy denying its economic origins (ie. an autonomy *outside* the domain of the only one practiced by the market participant who believed himself autonomous) that provided the bourgeois

> family with its consciousness of itself. (Jürgen Habermas, *The Structural Transformation of the Public Sphere: An Inquiry into a Category of Bourgeois Society*, trans. Thomas Burger [1962; Cambridge, MA: MIT Press, 1989], p. 46)

The popular representation of the "private sphere" of the family in the public form of the novel registers not merely the "privatization" of the process of reproduction (the family's individual economy) but also, and crucially, the ways in which the private sphere of society has become publicly relevant. If the everyday practices of the private sphere are allowed to appear in public—in the publicizing form of the novel, for instance—this is because the autonomy of the private sphere is itself a condition of its orientation toward the market itself. That is, the public relevance of the private sphere registers the necessary orientation of the "privatized" home toward the economic activities and commodity market that lie "outside" its confines and that, collaterally, are reinvented in the forms of "interiority" proper to it. Moreover, it is this coordination of interior states and external conditions, and the managing of the differences between private and public spheres, that forms part of the cultural work of the traditional novel itself. And it is precisely such a circulation and communication—or what amounts to a *commuting* between private and public spaces—that the midcentury public reading of the private (the romance) traces. (On the understanding of separation of spheres in Hawthorne in terms of the model, and the practice, of commuting, see my *Henry James and the Art of Power*, pp. 190–92.)

41. Chandler, *The Visible Hand*, pp. 27–28.
42. Leo Bersani, "The Jamesian Lie," in *A Future for Astyanax* (Boston: Little, Brown, 1976), p. 135.
43. The free styles of contractual connections available in James's reconsideration of the logic of contract are extensive. James's highly abstract and schematic treatment of models of marriage and family in the novel does not simply oppose Newman's new state of the family to the hereditary nobility of the Bellegardes (a succession of portraits in which Claire's "face was a larger and freer copy" of her mother's [p. 120] and Urbain "the old woman at second-hand" [p. 123]) but also involves the Babcock episode as a kind of homosocial trial marriage. As Newman puts it in his letter to Mrs. Tristram, "The nearest approach to [Claire] was a Unitarian minister from Boston, who very soon demanded a separation, for incompatibility of temper" (p. 76).
44. Thomas L. Haskell, "Capitalism and the Origins of the Humanitarian Sensibility, Part 2," *The American Historical Review* 90, 2 (June 1985), p. 551.
45. Haskell, "Capitalism and the Origins of the Humanitarian Sensibility,

Part 2," p. 553. On the relations of marriage, promise keeping, and contract in nineteenth-century America, see Michael Grossberg, *Governing the Hearth: Law and the Family in Nineteenth-Century America* (Chapel Hill: University of North Carolina Press, 1985), pp. 17–63. On contractualism generally, and particularly on the *restrictions* on contractualism by the later nineteenth century, see P.S. Atiyah, *The Rise and Fall of Freedom of Contract* (Oxford: Oxford University Press, 1979); see also, Morton J. Horwitz, *The Transformation of American Law* (Cambridge, MA: Harvard University Press, 1977).

46. Friedrich Nietzsche, *On the Genealogy of Morals* (New York: Vintage Books, 1969), pp. 57–58; cited by Haskell, "Capitalism and the Origins of the Humanitarian Sensibility, Part 2," pp. 551–52.
47. William James, *The Principle of Psychology*, vol. 1 (1890; repr. of first ed., New York: Dover, 1950), pp. 338–40. On possessive individualism, see C.B. Macpherson, *The Political Theory of Possessive Individualism: Hobbes to Locke* (Oxford: Oxford University Press, 1962), and, on the updating of the theory, Magali Sarfatti Larson, *The Rise of Professionalism: A Sociological Analysis* (Berkeley and Los Angeles: University of California Press, 1977), pp. 222–25. See also note 76, below.
48. On the fall of "radical" freedom of contract, see P.S. Atiyah, *The Rise and Fall of Freedom of Contract*. It should be clear that I am addressing the rise and fall of the model of contract and that practices and models cannot simply be conflated. My concern is with the "regression" to the logic of contract in machine culture and its affective and interpretive appeals.
49. Veblen, "The Place of Science in Modern Civilization," in *The Place of Science*, p. 30.
50. Adorno, "Veblen's Attack on Culture," pp. 86, 85.
51. Adorno, "Veblen's Attack on Culture," p. 81.
52. See Habermas, *The Structural Transformation of the Public Sphere*, p. 195. Habermas's account of the form this "refeudalization" takes ratifies Veblen's critique of the radiant body and its animistic status in the society of conspicuous consumption. For Habermas, "In the measure that it is shaped by public relations, the public sphere of civil society takes on feudal features. The 'suppliers' display a showy pomp before customers ready to follow. Publicity imitates the kind of aura proper to the personal prestige and supernatural authority once bestowed by the kind of publicity involved in representation" (p. 195). Veblen's hostility to consumer culture, is, of course, a hostility to personal prestige and representation both. On turn-of-the-century medievalism or refeudalization, see also T.J. Jackson Lears's treatment of "medieval mentalities in a modern world," *No Place of Grace*, pp. 141–81.

53. On the "physiognomy of goods," see Veblen, *Theory of the Leisure Class*, pp. 114–15.

54. See, for instance, in *The American*, pp. 17, 89, 188, 190.

55. If the identification of the female and the natural counters or allays this recognition, conserving the category of natural persons, it also has the effect of ratifying it, ultimately accounting natural persons—what Norris, for example, calls the "women people"—as facsimile or non-persons.

56. See, for example, Jacques Derrida, "Freud and the Scene of Writing," in *Writing and Difference*, trans. Alan Bass (Chicago: University of Chicago Press, 1978). For an astute account of the Derridean conceptualization of the scene of writing and writing-machines, see Jonathan Goldberg, *Writing Matter: From the Hands of the English Renaissance* (Stanford: Stanford University Press, 1990), particularly chap. 6 (pp. 282–318).

57. James's practice of composition in his later work—particularly, the practice of dictation—indicates one way in which James at once registers and manages such a radicalization of the materiality of writing and writing-technologies. I am referring, of course, to James's practice of dictating to a typewriter (the word referred originally to both the machine and its operator, usually a young woman). According to his typist, Mary Weld, James's dictation was "remarkably fluent" and "when working I was just part of the machinery." According to his more famous typist, Theodora Bosanquet, James wanted his typists to be "without a mind." Not surprisingly, Bosanquet, like William James, was interested in the psychophysics of "automatic writing." James thus reincorporates the automatisms of machine-writing in the practice of oral composition. That is, if the typewriter (as we have seen in the Introduction) *dis*articulates the links between mind, eye, hand, and paper, these links are *re*articulated in the dictatorial orality that "automatically" translates speech into writing. Hence James, responding to the suggestion that the typewriter and the practice of dictation affected his style of writing, insisted on the *transparency* or *im*materiality of such technologies of composition. As he put it, dictation to the machine "soon enough becomes *intellectually*, absolutely identical with the act of writing—or has become so, after five years now, with me; so that the difference is only material and illusory—only the difference that I walk up and down." (See Leon Edel, *The Master: 1901–1916* [New York: Avon Books, 1972], pp. 93–94, 360, 366, 127; Theodora Bosanquet, *Henry James at Work* [London: Hogarth Press, 1924].) Reducing technologies of writing to the "only" material and the material to the "illusory," James thus insists on the transparency of writing in general and on its disembodiment (such that, for instance, the difference of bodily motions—the difference of walking up and down—makes no difference).

But consider Theodora Bosanquet's comments on James's material practice of composition and the psychophysics of writing it registers: "Indeed, at the time when I began to work for him, he had reached a stage at which the click of a Remington machine acted as a positive spur. He found it more difficult to compose to the music of any other make. During a fortnight when the Remington was out of order he dictated to an Oliver typewriter with evident discomfort, and he found it almost disconcerting to speak to something that made no responsive sound at all" (p. 248). It is as if the "responsive sound"—the familiar "click" of the Remington—functions as the concerted response of an ideally responsive and automatized first reader, such that the "absolute identity" of the writing machine and the "act of writing" makes possible their mutual transparency (writing and registration immediately indicating each other) and hence makes material differences illusory differences.

The entire question of the *referentiality* of later nineteenth-century writing might be reconsidered in terms of such technologies of automatic and immediate *registration*. Hence the dream of perfect referentiality in realist writing (which I will discuss in the next part of this study, "Statistical Persons") is perhaps most productively to be considered in relation to such technologies of registration than reduced to the self-evidently dismissable desire, frequently attributed to realist writing, to ignore the medium of representation and to claim an unmediated access or "reference" to the real.

Practices of dictation, registration, and material impression, such as that of the typewriter key on paper or the spoken sound on the phonographic plate, cannot simply be reduced to instances among others of a writing in general ("abstracted" such that material differences in effect become illusory). Nor can scenes of writing simply be reduced to the technologies of writing that "determine" them ("materialized" in a technological determinism that in effect makes authorial intentions or individual differences illusory). It is not a matter, that is, of choosing between a general theory of writing, on the one side, or something like a new materialism, on the other. Again: things are, among other things, what they appear to be, and practices are neither simply reducible, nor simply irreducible, to anything else. If for James, for instance, the practice of dictation to the machine and the act of writing in general become "absolutely identical," it is *the work of making identical* (here, the work of making writing and mechanics equivalent such that each becomes transparent to the other) that must be analysed. In short: the writing of writing at the turn of the century locates, and specifies, the tension between the "material and illusory" or immaterial in terms of the body-machine complex. It is thus the relays by which new writing technologies and the relations of machine culture become reciprocally intelligible, and operational, that I am tracking here.

James, for example, powerfully registers the radical recompositions of writing and information-technologies at the turn of the century and in terms of what counts for James as "psychology," a psychology inseparable from the writing of writing. This is particularly evident in his fictions of the 1890s, registering the materiality of information-processing and technologies of communication (*In the Cage*, for example), the corporeality of thinking and speaking (*What Maisie Knew*, for example), the psychophysics, and pathologization, of reading and writing (*The Turn of the Screw*, for example). But James's dictatorial practice of writing, on these grounds at least, precisely obviates the conflations of the materiality of writing and technology visible, for instance, in Jack London's or Mark Twain's writing as working at the machine (see Introduction), or Frank Norris's attention to the mechanics of fiction writing (see Part I, "The Naturalist Machine"), or Stephen Crane's registration of the violent intersections of inscriptions, bodies, and machines (see Part III, "Statistical Persons"). What is most evident, particularly in James's earlier writings such as *The American*, is the pressured *re*articulation of writing and self-expression: that is, the *recovery* of the relations of identity, self-possession, and self-recognition that define, I have been arguing, the appeal of market culture and hence (I am suggesting) one way in which the regressive appeal to market relations allays the becoming visible of the machinelikeness of persons and writing in machine culture.

58. Marc Shell, "The Gold Bug," *Genre* 13, 1 (Spring 1980), 18. On the paper system, see Irwin Unger, *The Greenback Era: A Social and Political History of American Finance, 1865–1879* (Princeton: Princeton University Press, 1967); Richard Hofstadter, "Free Silver and the Mind of 'Coin' Harvey," in *The Paranoid Style in American Politics and Other Essays* (Chicago: University of Chicago Press, 1979), pp. 283–315; Walter T.K. Nugent, *Money and American Society, 1865–1880* (New York: Free Press, 1968). As Nugent observes in *The Money Question During Reconstruction* (New York: Norton, 1967), "No one realized in 1865, but money was destined to become the chief perennial issue in national politics for over thirty years . . . Its peculiar dimensions were established in almost all important ways during the Reconstruction years, from 1867–1879" (pp. 21–22).

59. George Eliot, *Middlemarch* (Harmondsworth: Penguin Books, 1965), p. 173.

60. See Ian Hacking, *The Taming of Chance* (Cambridge: Cambridge University Press, 1990), p. 105.

61. Ralph Waldo Emerson, "Fate" (1852), in *Selections from Ralph Waldo Emerson,* ed. Stephen E. Whicher (Boston: Houghton Mifflin, 1957), pp. 337, 333. On statistics and their social application, in addition to Hacking, *The Taming of Chance*, see Lorraine J. Daston, "Rational Individuals versus Laws of Society: From Probability to Statistics," in the indispensable collection, *The Probabilistic Revolution*, vol. 1, eds. Lorenz Kruger, Lorraine J. Daston, and Michael Heidelberger (Cambridge, MA: MIT Press, 1987), pp. 295–304. The making of statistical persons is the topic of the next part of this study.
62. Hacking, *The Taming of Chance*, p. 160.
63. Hacking, *The Taming of Chance*, p. 132.
64. The sense of action and agency I am very briefly and provisionally indicating here draws on and draws into relation, for example, H.L.A. Hart and Tony Honore, *Causation in the Law* (Oxford: Oxford University Press, 1985), particularly the discussions of the distinctions between causes and conditions; Gilbert Ryle, *The Concept of Mind* (New York: Barnes and Noble, 1959), particularly on motive and action; Michel Foucault, *Discipline and Punish*, on the linkages between the characterization and qualification of actors and the adjudication of actions, e.g., pp. 18, 247. My sense of these matters is less involved with the deconstruction of the possibility of "action," by way of a reduction of the "logic" of action or intention to its breaking point, than with the notion of action in, for example, some nonformalist versions of decision theory: decision, as *decidere* (literally, to cut off from the source) and action/decision as a necessary and non-invidious break in causal or logical sequence. (Here see, Niklas Luhmann, *Essays on Self-Reference* [New York: Columbia University Press, 1990] and the accounts of action and agency in Parts III and IV, below.) The resistance to the mixed or immanent account of agency I am sketching here is thus a resistance to the sort of resolutely undramatic accounts of action or decision set out in systems theory generally and an attempt to conserve what I've been calling the melodrama of uncertain agency.
65. On the liberal antinomies of possessive market society, see Roberto Mangabeira Unger, *Knowledge and Politics* (New York: The Free Press, 1975), esp. pp. 104–44.
66. See for instance, Stephen Greenblatt's brief article on "Culture," in the recent *Critical Terms for Literary Study*, eds. Frank Lentricchia and Thomas McLaughlin (Chicago: University of Chicago Press, 1990). Defining "culture" in terms of the antinomies of "mobility" and "constraint," and in terms of "commerce" and "exchange" in a "general economy," such a new historicist account in effect rewrites the liberal antinomies of market culture, and the vicissitudes of laissez-faire individ-

ualism, as the logic of culture *as such* (pp. 225–32). For a fundamental critique of the logicism of exchange that underlies such accounts, with particular reference to the cultural poetics of structural anthropology, see Pierre Bourdieu, *Outline of a Theory of Practice*, trans. Richard Nice (Cambridge: Cambridge University Press, 1977).

67. Thomas L. Haskell, "Capitalism and the Origins of the Humanitarian Sensibility, Part 1," *The American Historical Review* 90, 2 (April 1985), 341–53. On classical and modern conceptualizations of the category of "interest," see Patricia Springborg, *The Problem of Human Needs and the Critique of Civilization* (London: Allen & Unwin, 1981).

68. Foucault's remark, made in a personal communication, is cited in Hubert L. Dreyfus and Paul Rabinow, *Michel Foucault: Beyond Structuralism and Hermeneutics* (Chicago: University of Chicago Press, 1982), p. 187; Pierre Bourdieu, *Outline of a Theory of Practice*, p. 79.

69. Haskell, "Capitalism and the Origins of the Humanitarian Sensibility, Part 1," pp. 347, 343.

70. Veblen, *Theory of the Leisure Class*, pp. 184, 187, 188, 186.

71. I am thinking particularly about recent accounts of "structural causality" (Althusser), "structural intentionality" (Derrida), "objective intention" (Bourdieu), and also accounts of the irreducibility of structure and intention (de Man). See, for instance, Louis Althusser, *For Marx*, trans. Ben Brewster (London: New Left Books, 1977); Jacques Derrida, "Limited Inc," *Glyph 2: Johns Hopkins Textual Studies* (Baltimore: The Johns Hopkins University Press, 1977), pp. 192–217; Bourdieu, *Outline of a Theory of Practice*, pp. 78–81; Paul de Man, "The Purloined Ribbon," *Glyph 1: Johns Hopkins Textual Studies* (Baltimore: The Johns Hopkins University Press, 1977). Vincent Descombes provides an informed history of the shifting relations of subject and structure in *Modern French Philosophy*, trans. L. Scott-Fox and J.M. Harding (Cambridge: Cambridge University Press, 1980).

72. See Karl Polanyi, *Primitive, Archaic, and Modern Economics*, ed. George Dalton (New York: Doubleday, 1968) and *The Great Transformation* (New York: Holt, Rinehart, 1944).

73. Haskell, "Capitalism and the Origins of the Humanitarian Sensibility, Part 2," p. 547.

74. Walter Benn Michaels, "The Gold Standard and the Logic of Naturalism," *Representations* 9 (Winter 1985), 105–32. (Subsequently reprinted in *The Gold Standard and the Logic of Naturalism* [Berkeley: University of California Press, 1987].)

75. Compare, for instance, the argument of Michaels and Steven Knapp, "Against Theory," *Critical Inquiry* 8, 4 (Summer 1982), 723–42.

Whereas "Against Theory" argues for the inseparability of meaning and intention ("Meaning is just another word for expressed intention"), the double logic of "The Gold Standard" argues an irreducible distinction between intention and meaning and between "material and identity," and that double logic governs Michaels's account. My point is that such a double logic reproduces precisely the "radical formalism" (attributed here to de Man) that Michaels's argument against theory targets. Against theory in theory but reproducing its topics and techniques in practice, such new historicist work instances some of the ways in which the history/theory opposition has tended to function in that work. (See Fredric Jameson, *Postmodernism, or, The Cultural Logic of Late Capitalism* [Durham, NC: Duke University Press, 1991] for a related account of the way in which Michaels's reading of the market "rediscovers and reinvents" the "themes and issues" [p. 182] in recent theory and, more significant, a related account of the radical entanglement between the "against theory" argument—that one can never get outside beliefs—and "the gold standard" argument—that one can never get outside the market [pp. 211–17].)

The implications of these contradictions are clearest, for our purposes, in relation to the conflicts between market culture and machine culture I am setting out here. In an article called "Corporate Fictions" (reprinted in *The Gold Standard*), Michaels addresses the argument I make about persons and the body-machine complex in "The Naturalist Machine": as he puts it, "In fact, following Seltzer's lead, we can say that the 'discourse of force' not only undoes the opposition between body and machine but, perhaps more surprisingly, undoes the opposition between body/machine and the soul . . ." (p. 210). My problem with such a reading is not merely that Michaels's strictly logicist deconstruction proceeds too abstractly in "undoing" the differences between bodies and machines (and, therefore, eliminating, among other things, the problems of gender and differential cultural embodiment such differences "carry"). The larger problem is how quickly Michaels moves to recover exactly these oppositions and the notion of persons they conserve. Hence Michaels finds my description of the mechanics of naturalist writing and my appeal to Deleuze and Guattari's "notion of the 'desiring-machine' " "questionable," since, as he expresses it, "the basic point of that notion is to do away with persons, and its ultimate effect . . . is to do away with desire itself" (p. 210). But of course the basic point of Deleuze and Guattari's notion of the desire-machine is to insist on the relays between persons, desires, and machines, not their opposition. Michaels seems willing to risk such contradictions in order to protect the radical opposition of persons and desires to the machine process: that is, to conserve precisely the logic of possession and self-possession and the rituals of self-aggrandizement that, I have been arguing, define the regressions of market culture. Not the least remarkable contribution of Michaels's work is how powerfully and thoroughly it reproduces that logic and restages those rituals.

What I have set out here are some of the problems involved in the "anachronistic" (in the sense I've tried to specify) rewriting of disciplinary individualism and machine culture in the anterior terms of possessive individualism and market culture. As Deleuze and Guattari make clear, in their discussions of the "direct link . . . perceived between the machine and desire": "There is always something statistical in our loves, and something belonging to the laws of large numbers" (*Anti-Oedipus: Capitalism and Schizophrenia,* trans. Robert Hurley, Mark Seem, and Helen R. Lane [Minneapolis: University of Minnesota Press, 1983], pp. 285, 294). This "something statistical in our loves" is taken up, more directly, in the next part of this study, "Statistical Persons."

76. C.B. Macpherson, *The Political Theory of Possessive Individualism: Hobbes to Locke,* p. 273. For important reassessments of the theory of possessive individualism, see J.G.A. Pocock, *Virtue, Commerce, and History: Essays on Political Thought and History, Chiefly in the Eighteenth Century* (Cambridge: Cambridge University Press, 1985), esp. chaps. 3 and 6; Elizabeth Fox-Genovese and Eugene D. Genovese, *Fruits of Merchant Capital: Slavery and Bourgeois Property in the Rise and Expansion of Capitalism* (Oxford: Oxford University Press, 1983), esp. chap. 10, "Physiocracy and Propertied Individualism."

77. Foucault, *Discipline and Punish*, p. 222. See also Foucault's following discussion of the disciplines "as a sort of counter-law":

> discipline creates between individuals a "private" link, which is a relation of constraints entirely different from contractual obligation; the acceptance of a discipline may be underwritten by contract; the way in which it is imposed, the mechanisms it brings into play, the non-reversible subordination of one group of people by another, the "surplus" power that is always fixed on the same side, the inequality of position of the different "partners" in relation to the common regulation, all these distinguish the disciplinary link from the contractual link . . . whereas the juridical systems define juridical subjects according to universal norms, the disciplines characterize, classify, specialize; they distribute along a scale, around a norm, hierarchize individuals in relation to one another, and, if necessary, disqualify and invalidate . . . in the genealogy of modern soci-

> ety, [the disciplines] have been, with the class domination that traverses it, the political counterpart of the juridical norms according to which power was redistributed. (p. 223)

Hence the "anachronistic" rewriting of the disciplines of machine culture in terms of the model of contract and logic of market culture supports the legal fictions of "modern society." I will return to the functional opposition between corporal disciplines and formal principles in the final part of this study.

78. The capitalist, according to Marx, must "have fine hearing and a thick skin; must be simultaneously cautious and venturesome, a swashbuckler and a calculator, careless and prudent. He must, in fine, develop all the qualities of an experienced man of business." Marx, as quoted in Matthew Josephson, *The Robber Barons: The Great American Capitalists, 1861–1901* (New York: Harcourt, Brace, 1934), p. 192.

79. See my "*The Princess Casamassima*: Realism and the Fantasy of Surveillance," *Nineteenth-Century Fiction* 35, 4 (March 1981), 506–34; repr. in *Henry James and the Art of Power*.

Part III: Statistical Persons

1. All references to *Maggie* are to the Norton Critical Edition of *Maggie: A Girl of the Streets*, ed. Thomas A. Gullason (New York: Norton, 1979), which reprints the 1893 edition. All other references to Crane's work are to the Library of America edition *Stephen Crane: Prose and Poetry* (New York, 1984). Subsequent references to both are included parenthetically in the text.

2. I am concerned here primarily with the American "realist" and "naturalist" fiction of the 1880s and 1890s. I am less interested for the moment in the differences, or differences in emphasis, between these narrative kinds than in the network of relations they have in common. I would suggest, however, that the normative distinctions made between these kinds (most emphatically, the Lukácsian discrimination between realism and naturalism) tend to register what I will be considering as the internal tensions and oppositions that structure both only as an opposition between them.

3. Samuel Langhorne Clemens (Mark Twain), *A Connecticut Yankee in King Arthur's Court*, ed. Allison R. Ensor (New York: Norton, 1982); subsequent page references are included parenthetically in the text.

4. I am here drawing on, and extending the terms of, the account of the realist investment in vision and supervision set out in my "*The Princess*

Casamassima: Realism and the Fantasy of Surveillance," *Nineteenth-Century Fiction* 35, 4 (March 1981), 506–34; repr. in my *Henry James and the Art of Power* (Ithaca: Cornell University Press, 1984).

5. I here adapt the terms of Michael Fried's *Absorption and Theatricality: Painting and Beholder in the Age of Diderot* (Berkeley: University of California Press, 1980).

6. On what I have elsewhere called the "rule of reversibility" in realist representations of power, see my *Henry James and the Art of Power*, esp. chap. 3.

7. Michel Foucault, *Discipline and Punish: The Birth of the Prison*, trans. Alan Sheridan (New York: Pantheon, 1977), p. 278.

8. Sigmund Freud, "On Narcissism," in *The Standard Edition of The Complete Psychological Works of Sigmund Freud*, ed. James Strachey (London: Hogarth, 1953–74), 14:89.

9. Naomi Schor, *Reading in Detail: Aesthetics and the Feminine* (New York and London: Methuen, 1987), p. 38.

10. Crane, in his notorious defense of the alleged prostitute Dora Clarke from police harassment, not at all atypically experienced the working of this recidivist machinery; Crane experienced, as R. W. Stallman observes, "the corrupt power of New York City's policemen, who tolerated can-can and peep shows and houses of prostitution so long as they remained within a restricted [and well-policed] area." *Stephen Crane: A Biography* (New York: George Braziller, 1968), p. 221. On the policing of prostitution in the nineteenth century, see Judith Walkowitz, *Prostitution and Victorian Society: Women, Class and the State* (Cambridge: Cambridge University Press, 1980); Alain Corbin, *Les filles de noce: Misère sexuelle et prostitution aux 19ième et 20ième siècles* (Paris: Aubier Montaigne, 1978); and Barbara Meil Hobson, *Uneasy Virtue: The Politics of Prostitution and the American Reform Tradition* (New York: Basic, 1987). My interest here, it should be clear, is less in the social history of prostitution than in the ways in which relations of visibility, representation, and embodiment make it possible for the figure of the prostitute and the problem of "the social" to indicate each other, across a range of discourses, practices, and institutions of regulation.

11. Such conflations of the sexual, the familial, and the social (what Crane in *Maggie* calls "the home street") provide the subject and method of Crane's later story of the slums, *George's Mother* (1895). The policing of families in the exploratory and reformist literature on the urban underworld has been well documented. It might be epitomized by the exchange recorded by the reformer Thomas De Witt Talmadge, in *The Night Sides of City Life* (Chicago: J. Fairbanks, 1878): " 'My dear boys, when your father and your

mother forsake you, who will take you up?' They shouted, 'The perlice, sir; the perlice?' " (p. 134). On the significance of the reformist writings of Crane's father Jonathan Townley Crane for the novelist's treatment of the slums, see Thomas A. Gullason, "The Sources of Stephen Crane's *Maggie*," *Philological Quarterly* 38 (1959), 497–502.

12. As Theodore M. Porter documents, in his *The Rise of Statistical Thinking 1820–1900* (Princeton: Princeton University Press, 1986): "Without detailed records, centralized administration is almost inconceivable, and numerical tabulation has long been recognized as an especially convenient form for certain kinds of information," p. 17. The account of statistical persons I am setting out here is indebted to, in addition to sources cited below: Porter (esp. pp. 149–319); Lorraine J. Daston, "Rational Individuals versus Laws of Society: From Probability to Statistics" and Ian Hacking, "Prussian Numbers 1860–1882," among other essays in the excellent collection *The Probabilistic Revolution*, 2 vols., eds. Lorenz Kruger, Lorraine J. Daston, and Michael Heidelberger (Cambridge, MA: MIT Press, 1987); Ian Hacking, *The Emergence of Probability* (Cambridge: Cambridge University Press, 1975), "Making Up People," in *Reconstructing Individualism: Autonomy, Individuality, and the Self in Western Thought*, eds. Thomas C. Heller, Morton Sosna, and David E. Wellbery (Stanford: Stanford University Press, 1986), "Biopower and the Avalanche of Printed Numbers," *Humanities in Society* 5 (1982); and Stephen M. Stigler, *The History of Statistics: The Measurement of Uncertainty before 1900* (Cambridge, MA: Harvard University Press, 1986).

13. See Allan Sekula, "The Body and the Archive," *October* 39 (1986), 3–64; on the nexus of surveillance and statistics in the literature of the urban underworld, see my "*The Princess Casamassima*: Realism and the Fantasy of Surveillance."

14. See: Francis Galton, "On Generic Images," *Proceedings of the Royal Institution* 9 (August 2, 1890), 166; Galton, "Generic Images," *Nineteenth Century* 6, 29 (July 1879), 162–69; and Galton, *Inquiries into the Human Faculty and Its Development* (London: Macmillan, 1883), pp. 5–6, 17. On generic individuality, see also Carlo Ginzburg, "Morelli, Freud, and Sherlock Holmes: Clues and Scientific Method," *History Workshop* 9 (Spring 1980), 5–29. On generic pictures, see also: Edward R. Tufte, *The Visual Display of Quantitative Information* (Cheshire, Conn.: Graphics Press, 1983). I will return to the problem of pictorial statistics and other instances of what might be called "working models" in the last section of this chapter.

15. Jacob Riis, *How the Other Half Lives: Studies Among the Tenements of New York* (1890) (repr. of 1901 ed., New York: Dover, 1971), p. 205.

16. I will return to some of these contemporary instances in the chapter that

follows, with particular reference to the ways in which a good deal of current cultural criticism continues to inhabit the logistics of realism.

17. Elaine Scarry, *The Body in Pain: The Making and Unmaking of the World* (New York: Oxford University Press, 1985). Subsequent references are included parenthetically in the text.

18. Scarry's version of "transcendental realism" (and I refer here both to her argument and to the imitative, narrative, and even "incarnationist" form of that argument) inhabits the terms of the double discourse (the relays between the visible and the corporeal) we are considering, but rewrites these terms in the form of a radical opposition. This opposition is protected through an ethics of "carefully controlled" referentiality. Hence Scarry's ethical realism depends on an exact "correspondence between language and material reality" (p. 10). The imperative is to "hold steadily visible the referent" (p. 17) in order to guarantee the utter transparency of language to its referents: what amounts to "the identification of the physical and verbal acts." The coming apart of this identity, through a "perceptual confusion sponsored by the sign," opens the possibility of what Scarry describes as "fiction." Hence ethical realism, guarding against the confusion sponsored by the sign, requires the "obligatory referentiality of fiction" (p. 258). By this account, "power" is always a "fiction of power" (e.g. p. 27) that fraudulently interrupts this referential obligation. What the identification of language and material reality, of physical and verbal acts, makes possible (and what the confusion of power interrupts) is thus the translation of the physical into the verbal and the transcendence of material reality in language. Scarry's incarnationist rhetoric everywhere has the double effect of embodying abstractions and abstracting the body. One effect of this abstract and disembodied realism is to identify the body with pain and the cultural work of embodiment as the transcendence both of the body and pain. The body and culture are thus "mutually exclusive" and the cultural work of embodiment must "refer back" to the body but only in order to enable the radical disembodiment of persons and personhood. If, as Scarry puts it, "every act of civilization is an act of transcending the body in a way consonant with the body's need" (p. 57), such an avowed honoring of the body's need covers for its abstraction, since the body's need is precisely the need to be transcended. Thus, even sexual pleasure is here presented as an experience of utter disembodiment (p. 166). My sense is that what I have been calling Scarry's version of transcendental realism depends on an untenable notion of linguistic regulation and reference and on an idealized notion of the transparency of signs to referents and bodies to language: not surprisingly, "deconstruction" is the same as destruction, in Scarry's idiom. But, whatever the difficulties with the policing of reference Scarry's

story of language involves, this version of transcendental realism has the effect of reconstructing and conserving what might be called *the romance of culture* (a romance that, as I have noted in the previous chapter, is, more specifically, a romance of liberal market culture). One effect of this transcendental realism and romance of culture, for example, appears in the recurrent tendency in this text to oppose stories of third world torture and pain to lyrical accounts of the domestic intimacies of the "civilized" world (e.g. p. 40)—an opposition that at times takes the form of the familiar "political" opposition of lower and upper halves as the opposition of body to soul. It is perhaps also this opposition that is policed and defended by what Scarry describes as "the imagination . . . constantly at work, patrolling the dikes of made culture . . . like a watchman patrolling the dikes of culture by day and by night" (pp. 321, 325). The naturalization of the threat to the "dikes of culture" and the transcendentalizing of the police-work of border defense (the "imagination" patrolling the dikes) seems to naturalize and transcendentalize an artifact-rich and thoroughly "made culture" set melodramatically against the overly embodied and underly artifactualized enemy that threatens to overwhelm it. In short, Scarry's narrative registers the realist entanglement between the body and its embodiments but only in terms of an essential opposition: only by positing a well-regulated referentiality, immune to the perceptual confusions sponsored by the sign (culture) or as a sign of that confusion (power). What is therefore disavowed (naturalized or transcendentalized) is the possibility of the *internal* relations between culture and power: that is, between the "made" world and the resolutely less than, or more than, abstract bodies, cultures, and powers that remain outside the romance of culture.

19. Foucault, *Discipline and Punish*, pp. 156, 203.
20. Foucault, *Discipline and Punish*, p. 177.
21. George Eliot, *Daniel Deronda*, ed. Barbara Hardy (Harmondsworth: Penguin, 1967); Edith Wharton, *The House of Mirth* (Boston: Houghton Mifflin, 1963). Subsequent references are to these editions and are included parenthetically in the text.
22. On the capitalizing on risk in Wharton's novel, see Walter Benn Michaels, *The Gold Standard and the Logic of Naturalism* (Berkeley: University of California Press, 1987), pp. 217–44. Whereas Michaels reincorporates risk and chance within a general logic of market culture, for me Wharton's novel measures the differences and tensions between market culture and machine culture and the limits of the logic of "the market"; and whereas Michaels (in response to the argument about persons and mechanisms set out in "The Naturalist Machine") conserves a "basic" opposition between desire and the machine (*The Gold Standard*, pp. 209–10, and

see above, p. 199n75), for me it is precisely the internal relations between desire and the machine, the something mechanical or statistical in persons that incites desire, that becomes visible in these cases.

23. William James, *The Principles of Psychology* (Cambridge, MA: Harvard University Press, 1983), pp. 19, 143.
24. My comments here on chance and determinism in psychoanalysis are directly indebted to Jacques Derrida's "My Chances/*Mes Chances*: A Rendezvous with Some Epicurean Stereophanies," in *Taking Chances: Derrida, Psychoanalysis, and Literature*, eds. Joseph H. Smith and William Kerrigan (Baltimore: The Johns Hopkins University Press, 1984).
25. Sigmund Freud, *The Psychopathology of Everyday Life*, as cited in Derrida, "Taking Chances/*Mes Chances*," p. 23.
26. On Laplace, see Theodore M. Porter, *The Rise of Statistical Thinking 1820–1900*, p. 51.
27. In several senses, because the word "case" derives from and remains linked to "casus" or fall; and the fallen person, the prostrate body, is the paradigmatic case of falling ill or out of luck. See Derrida, "Taking Chances/*Mes Chances*," pp. 4–7. Pertinent here too is the longstanding link between statistics and fatality, the statistical fatality that underwrites, for example, the centering of early statistical sociology (anticipated in the case of Durkheim, for instance) on the problem of suicide. On statistical fatality, see Ian Hacking's superb *The Taming of Chance* (Cambridge: Cambridge University Press, 1990), pp. 114–31.
28. Michael Fried, *Realism, Writing, Disfiguration: On Thomas Eakins and Stephen Crane* (Chicago: University of Chicago Press, 1987).
29. See John Berryman, *Stephen Crane* (New York: Sloane, 1950).
30. Something more, that is, than the tautological relations between self-absorption and the act of writing (such that the scene of writing indicates intense self-absorption which indicates the scene of writing) conserved in accounts such as Fried's, or in other quasi-phenomenological and abstracted versions of deconstruction. The writing of writing at the turn-of-the-century, as Friedrich Kittler has traced, does not conform to the continuum of writing and self-identity articulated in the "romantic" discourse network of 1800. And the insistent reduction of technologies of writing to a problem of identity or to the vicissitudes of "the self" is one way of conserving (albeit in excruciated form—in the form of "mourning") a romance of writing in machine culture: one way of recuperating the romance of writing, and its appeals, in the "naturalist" discourse network of 1900. See Friedrich Kittler, *Discourse Networks, 1800/1900*, trans. Michael Metteer, with Chris Cullens (Stanford: Stanford University Press, 1990), pp. 177–212. See also the more extended discussion of naturalist writing-technologies in my Introduction.

31. Donald Davidson, *Essays on Actions and Events* (Oxford: Oxford University Press, 1985), p. 121.
32. Eliot, *Daniel Deronda*, p. 470.
33. Charles Sanders Peirce, "The Doctrine of Chances," *Popular Science Monthly* 12 (1878), excerpted in *The World of Laws and the World of Chance*, ed. James R. Newman (New York: Simon and Schuster, 1956), pp. 1338–39. On Peirce and "a universe of chance," see Ian Hacking, *The Taming of Chance*, pp. 200–15; Peirce, concludes Hacking, was "a man whose professional life as a measurer was immersed in the technologies of chance and probability, and who, in consequence of that daily experience, finally surrendered to the idea that there is absolute chance in the universe," p. 215.
34. On the problem of the body in Melville's *Moby-Dick*, see Sharon Cameron's stimulating *The Corporeal Self: Allegories of the Body in Hawthorne and Melville* (Baltimore: The Johns Hopkins University Press, 1981), pp. 23–24.
35. I am indebted to Michael Fried for mentioning to me the coincidence of these articles and for suggesting its pertinence to my argument. Fried briefly comments on "In the Depths of a Coal Mine" in his *Realism, Writing, Disfiguration* (pp. 137, 140), stating that figures such as the engineer in Crane's article are "instantly recognizable as surrogates for the writer" (p. 140). But here again this seems too instantly to *read through* the specific instances and medium of Crane's writings; as we will see in a moment, such a reading of the engineer is only the case if "the writer" is understood as a function of the control and registration technologies that here enclose him.
36. George Santayana, *Character & Opinion in the United States* (New York: Scribner's, 1930), p. 175. On the machine as mediator between idealism and materialism, as the "operator" of transcendental materialism, see also Michael Pupin, *The Romance of the Machine* (New York: Scribner's, 1930): "the machine is the visible evidence of the close union between man and the spirit of the eternal truth which guides the subtle hand of nature" (pp. 28–29).

Part IV: The Still Life

An earlier version of this paper was circulated and presented at the Center for Literary and Cultural Studies, Harvard University, March 1989. I am indebted to the seminar participants at the Center, and to audiences and colleagues at

Boston College, University of California at Santa Cruz, Tufts University, and Cornell University for their stimulating and instructive responses to the piece.

1. Jean Baudrillard, *For a Critique of the Political Economy of the Sign*, trans. Charles Levin (St. Louis: Telos Press, 1981), p. 85.
2. See Catherine Gallagher, "The Bio-Economics of *Our Mutual Friend*," in *Fragments for a History of the Human Body*, Part Three, ed. Michel Feher, Ramona Naddaff, and Nadia Tazi. [Zone 5] (New York: Urzone, 1989), 345–65.
3. The quoted passage is from Jean Laplanche, *Life and Death in Psychoanalysis*, trans. Jeffrey Mehlman (Baltimore: The Johns Hopkins University Press, 1976), p. 102. For an incisive account of the pornography question, see Susan Stewart, "The Marquis de Meese," *Critical Inquiry* 15, 1 (Autumn 1988), 162–92.
4. Thorstein Veblen, "The Economic Theory of Woman's Dress," *The Popular Science Monthly* 46 (December 1894), 198–205; subsequent references to this article are included parenthetically in the text.
5. Elizabeth Wilson, *Adorned in Dreams: Fashion and Modernity* (Berkeley and Los Angeles: University of California Press, 1985), p. 127.
6. Georg Simmel, "Fashion," (originally published in *International Quarterly* [New York, 1904]); repr. in *On Individuality and Social Forms: Selected Writings*, ed. Donald N. Levine (Chicago: University of Chicago Press, 1971), pp. 294–96.
7. Daniel Boorstin, *The Americans: The Democratic Experience* (New York: Random House, 1973), pp. 89–244.
8. This is to reconsider both the pejorative understanding of "weightlessness" as an effect of consumer culture (for instance, in the account of turn-of-the century American culture provided by T.J. Jackson Lears) and the euphoric understanding of the artifactuality of persons in consumption (for instance, what I take to be the romance of market culture implicit in Elaine Scarry's account of bodies and artifacts). See T.J. Jackson Lears, *No Place of Grace: Antimodernism and the Transformation of American Culture, 1880–1920* (New York: Pantheon Books, 1981), pp. 32, 42–47; Elaine Scarry, *The Body in Pain: The Making and Unmaking of the World* (New York: Oxford University Press, 1985). On the assumption of "bourgeois acquisitiveness" in Scarry's work, see Jonathan Goldberg, *Writing Matter: From the Hands of the English Renaissance* (Stanford: Stanford University Press, 1990), pp. 314–16; and, for a somewhat different account of Scarry's work and of the idealizations of market culture in recent body-talk and agency-talk generally, see Parts II and III, above.

9. Rebecca Harding Davis, *Margret Howth: A Story of To-day* (Boston: Ticknor and Fields, 1862). Subsequent page references are included parenthetically in the text.
10. See Elizabeth Wilson, *Adorned in Dreams*, p. 2.
11. See, for instance, Gerald Heard, *Narcissus: Anatomy of Clothes* (London: Kegan Paul, 1924). Commenting on the "evolution" of the anatomy of clothes, Heard argues for the radical attenuation of the natural body: "our bodies," as he puts it, "may be on the way to disappear" (p. 155). On consumerism and the anorexic ideal, see Stuart Ewen, *All Consuming Images: The Politics of Style in Contemporary Culture* (New York: Basic Books, 1988), p. 183. For a detailed account of the problem of the body in market culture, see Part II, "Physical Capital: The Romance of the Market in Machine Culture."
12. On Loeb and the engineering of the life process, see Philip J. Pauly, *Controlling Life: Jacques Loeb and the Engineering Ideal in Biology* (New York: Oxford University Press, 1987). For a more sustained look at the topics of culturalism, naturalism, and the melodrama of uncertain agency, see Part V, "The Love-Master."
13. A good deal of even the most powerful recent cultural criticism (and here I have in mind the strictly logicist deconstructions that have marked some new historicist accounts of the market) has tended to idealize such a logic, or tautologic, of equivalence, while eliding the *logistics*, cultural and social, of making-equivalent. It is such a conflation of the things of logic and the logic of things that, for example, leads Marx to describe logic as "the money of the mind." (See note 25, below).
14. As I will be tracing in the next chapter, the rule-of-thumb that has guided much recent criticism might be restated in these terms: When confronted by the nature/culture opposition, choose the culture side. What has sponsored this preference (and its virtually automatic anti-naturalism) are the political advantages that seem to accrue from it: If persons and things are constructed, they could (at least in principle) be constructed differently. But such an abstract preference for the constructed, like the abstract preference for "difference" in a certain style of deconstructive criticism, has produced basic if artificial dilemmas, not least the apparent elimination of any possibility of a politics at all. For one thing, if everything could *in principle* be different, this scarcely tells us anything about what *in practice* is more or less recalcitrant to change. For another, and again like the abstract preference for difference, the insistence on the "arbitrariness" of persons and things has been taken to indicate the impossibility of claiming to act *in principle* at all. (Hence the recourse to emergency notions such as "strategic essentialism," in Gayatri Spivak's formulation, in order to restore at least tactical grounds for principles and preferences.)

Put simply, this critical work proceeds as if the deconstruction of the traditional dichotomy of the natural and the cultural (or of production and consumption, or of use and exchange) indicated merely the elimination of the first term and inflation of the second. Rather than mapping how the relays between what counts as natural and what counts as cultural are differentially articulated, invested, and regulated, and rather than challenging the terms of the nature/culture antinomy and the account of agency that antinomy entails, the tendency has been to discover again and again that what seemed to be natural is in fact cultural. (On the choice of the culture side and its implications, see Part V, which places the argument I make here in a somewhat different context.)

15. Bryan S. Turner, *The Body and Society: Explorations in Social Theory* (Oxford and New York: Basil Blackwell, 1984), p. 2. For a critique of such a logic of identity, see Denise Riley, *"Am I that Name"? Feminism and the Category of "Women" in History* (Minneapolis: University of Minnesota Press, 1988).

16. See Part III: "Statistical Persons."

17. Wakefield, as cited by Sidney W. Mintz, *Sweetness and Power: The Place of Sugar in Modern History* (New York: Penguin, 1985), pp. 253–54. As Mintz adds, "so much for symbolic anthropology."

18. Joan Copjec, "Cutting Up," in *Between Feminism and Psychoanalysis*, ed. Theresa Brennan (New York: Routledge, 1989), pp. 227–43. Cf. also Jean Baudrillard, *For a Critique of the Political Economy of the Sign*: "We do not wish to say that 'the individual is a product of society' at all. For, as it is currently understood, this culturalist platitude only masks the more radical truth that, in its totalitarian logic, a system of productivist growth (capitalist, but not exclusively) can only produce and reproduce men—even in their deepest determinations: in their liberty, in their needs, in their very unconscious—as productive forces. The system can only produce and reproduce individuals as elements of the system" (p. 86).

19. Edward A. Filene, as cited by Christopher Lasch, *The Culture of Narcissism: American Life in an Age of Diminishing Expectations* (New York: Norton, 1978), pp. 71–72.

20. Rebecca Harding Davis, *Life in the Iron Mills; Or, The Korl-Woman* (Old Westbury, NY: Feminist Press, 1972), p. 11.

21. See Henry Ford, *My Life and Work* (Garden City, NJ: Doubleday, Page, 1923); King Camp Gillette, *The People's Corporation* (Boston: Ball, 1924), p. 152.

22. Standard accounts of industrial standardization, and the relocation of the relations between heads and hands in the work process, include Harry Braverman, *Labor and Monopoly Capital: The Degradation of Work in*

the Twentieth Century (New York: Monthly Review Press, 1974), pp. 58–138; David F. Noble, *America by Design: Science, Technology, and the Rise of Corporate Capitalism* (Oxford: Oxford University Press, 1977), pp. 257–320. For a provocative account of the reimagination of the work process and the working body in the later nineteenth century, see Anson Rabinbach's indispensable *The Human Motor: Energy, Fatigue, and the Origins of Modernity* (New York: Basic Books, 1990).

23. Michel Foucault, *Discipline and Punish: The Birth of the Prison* (New York: Pantheon, 1977), p. 194. For a richly illustrated account of changing notions of individuality in the nineteenth century, see Alain Corbin, "The Secret of the Individual," in *A History of Private Life*, vol. 4 (Cambridge, MA: Harvard University Press, 1990), pp. 457–547.

24. Charles Sanders Peirce, *Collected Papers*, eds. Charles Hartshorne and Paul Weis (Cambridge, MA: Harvard University Press, 1931), 11: 159.

25. There are of course significant variants of this "post-oppositional" or consensus criticism (the names here would include Sacvan Bercovitch, Philip Fisher, Walter Benn Michaels, and Donald Pease), work epitomized in part by the contributions to the forthcoming Cambridge American Literary History. But the recent and general "reconstruction" of American literary history, a reconstruction along the newly renovated lines of the ineluctibility of the liberal consensus and, in effect, the inevitability of market relations, is scarcely limited to these instances. Such "cultural work" has come to represent a sort of American studies "antidote" to recent work in cultural studies and feminist studies. It also represents, at least in some quarters, what might be called the "gentrification" of the new historicism.

26. Niklas Luhmann, *Essays on Self-Reference* (New York: Columbia University Press, 1990), p. 8.

27. I have in mind, for instance, the "de Manian" accounts of the lethal effects of the aporetic dislocation of logical and temporal determinations (the "unavoidable moment" of arbitrariness in action). I am not disputing such a logical/temporal disruption. I am disputing instead the melodramatic consequences, and the resolutely abstract "politics" of signification, that are drawn from it. (The attempt to derive a politics from the logical and temporal unfoundedness of actions and decisions at once idealizes the conditions of action and decision and guarantees the rehearsals of the "impossibility" of action—and hence guarantees the rehearsals of what I have been calling the melodrama of uncertain agency.) I have been suggesting some of the ways in which such accounts of action and intention, and such instances of *agency-talk*, lend themselves to an idealization of the market criteria of self-possession and self-identity and in effect conserve, albeit in negative form, an understanding of action

and intention in terms of the logic of possessive individualism and personhood in market culture. On these points, see the discussions of action and agency-talk in Part II ("Physical Capital") and Part III ("Statistical Persons").

28. Jacques Lacan, *The Four Fundamental Concepts of Psycho-Analysis*, trans. Alan Sheridan (New York: W.W. Norton, 1977), p. 81.
29. Alfred Sohn-Rethel, *Intellectual and Manual Labor: A Critique of Epistemology* (Atlantic Highlands, NJ: Humanities Press, 1978), p. 45.
30. See C.B. Macpherson, *The Political Theory of Possessive Individualism: Hobbes to Locke* (Oxford: Oxford University Press, 1962).
31. Emile Zola, *La Ventre de Paris* (1873), *Oeuvres completes*, vol. 4 (Paris: E. Fasquelle, 1927), pp. 29, 335.
32. On the unnaturalness of Nature in naturalism, made visible in such representations, see Part V: "The Love-Master."
33. See Meyer Schapiro, "The Apples of Cézanne: An Essay on the Meaning of Still-life," in *Modern Art 19th and 20th Centuries: Selected Papers* (New York: George Braziller, 1978), p. 29. I can here merely broach the network of relations that traverses the still life. In addition to Lacan and Schapiro, this brief account is informed by these very different approaches to the still life: Jean Baudrillard, "The Trompe-L'oeil," in *Calligram: Essays in the New Art History from France* (Cambridge: Cambridge University Press, 1988), pp. 53–62; Carol Armstrong, "Reflections on the Mirror: Painting, Photography, and the Self-Portraits of Edgar Degas," *Representations* 22 (Spring 1988), 108–41; Françoise Meltzer, *Salome and the Dance of Writing* (Chicago: University of Chicago Press, 1987); Norman Bryson, "Chardin and the Text of Still Life," *Critical Inquiry* 15, 2 (Winter 1989), 227–52.
34. The still life thus excludes persons at the same time as it everywhere includes proxies for persons. One might list the ascending status of these proxies, and their share of the properties of persons, extending from representations of fruit or game to insects to birds to monkeys to statues or facsimile persons. The still life thus allocates personhood according to a principle of scarcity while accruing signs of personhood in the properties of surrogates, for example, in the sheer motility of insects or in the lure of representations that attract birds or monkeys. The difference between representation as the lure of appearance and representation as indicating the something beneath appearance that gives the appearance, is, as we will see in a moment, one way in which the still life ratifies the difference between persons and things that resemble persons.
35. See, for instance, John Berger, *About Looking* (New York: Pantheon, 1980); Svetlana Alpers, *Rembrandt's Enterprise: The Studio and the Market* (Chicago: University of Chicago Press, 1988).

36. Georg Lukács, "Narrate or Describe?," in *Writer and Critic and Other Essays*, ed. and trans. Arthur Kahn (London: Merlin Press, 1970), pp. 131, 138–46.

37. Max Horkheimer and Theodor W. Adorno, *The Dialectic of Enlightenment* (New York: Herder and Herder, 1972), p. 167.

38. We may note here one of the basic problems with the demystification model of ideology critique or with the notion that "exposing" the natural as the cultural is the most useful form of that critique, at least with reference to consumer society. What appears as scandalous in accounts of the culture industry is that consumers buy things and representations even though "they see through them." But it might be argued that the aesthetics of consumption, and that the pleasure in illusion that consumerism makes more generally and more indiscriminately possible, works rather differently. P.T. Barnum, for example, recognized early on that his success depended less on the credulity of his public than on its suspicion. That is, the cultivated egregiousness of his promotional schemes explicitly incited his paying public to "see through" the mechanisms of deception and illusion, to see through the humbug. This is the democracy of disillusionment in a republic where each citizen-consumer can and will pay to see through the cover of rationality to the cleverness of its mechanisms of illusion: in such a democracy everyone will pay to see that the Emperor has no clothes. The mechanisms of such a disillusionment-machine easily slip through the demystification model of ideology critique. (On Barnum, see Neil Harris, *Humbug: The Art of P.T. Barnum* [Boston: Little, Brown, 1973] and A.H. Saxon, *P.T. Barnum: The Legend and the Man* [New York: Columbia University Press, 1989].)

39. Lacan, *The Four Fundamental Concepts*, p. 112.

40. Lacan, *The Four Fundamental Concepts*, p. 107. Lacan's comments on the attraction of illusionism refer back to the paradigmatic example of illusionism in the Western tradition, the contest between Zeuxis and Parrhasios, relayed by Pliny: "The story runs that Parrhasios and Zeuxis entered into competition, Zeuxis exhibiting a picture of some grapes, so true to nature that the birds flew up to the wall of the stage. Parrhasios then displayed a picture of a linen curtain, realist to such a degree that Zeuxis, elated by the verdict of the birds, cried out that now at last his rival must draw the curtain and show his picture. On discovering his mistake he surrendered his prize to Parrhasios, admitting candidly that he had deceived the birds, while Parrhasios had deluded himself, a painter." *The Elder Pliny's Chapters on the History of Art* (London: Macmillan, 1890), pp. 109–10. As Lacan glosses it, the contest illustrates the difference between "the natural function of the lure and that of *trompe-l'oeil*": "if one wishes to deceive a man, what one presents to

him is the painting of a veil, that is to say, something that incites him to ask what is behind it" (p. 112). On some of the implications of this episode for a reading of the relations of gender and illusionism, see Naomi Schor, *Reading in Detail: Aesthetics and the Feminine* (New York and London: Methuen, 1987), pp. 12–14, 149–50.

41. This is, as Georg Simmel somewhat differently puts it, the something beneath appearances that constitutes "a synthesis of the individual's having and being." As Simmel expresses it, commenting on the attraction of adornment, "the attraction of the 'genuine,' in all contexts, consists in its being more than its immediate appearance, which it shares with its imitation." Instancing the difference between the genuine and the imitation, Simmel continues: "The 'genuine' individual, thus, is the person on whom one can rely even when he is out of one's sight. In the case of jewelry, this more-than-appearance is its *value*, which cannot be guessed by being looked at, but is something that . . . is *added* to the appearance." *The Sociology of Georg Simmel* (New York: The Free Press, 1950), pp. 338–44. It is therefore not quite the more-than-appearance that makes up the value of the genuine but the difference between the appearance and the more-than-appearance: the attraction of the genuine jewel, like the attraction of the genuine person, is that you can't tell merely by looking at it whether it's genuine or not. And the attraction of taking possession is, for Simmel as for Lacan, the ratification of the difference between the appearance and the more-than-appearance that ratifies the certainty of being a subject.

42. Ingvar Bergstrom, *Dutch Still-Life Painting in the Seventeenth Century* (London: Faber and Faber, 1956), p. 3.

43. Sohn-Rethel, *Intellectual and Manual Labor*, p. 25.

44. On motion and consumption (with specific reference to Zola), see Wolfgang Schivelbusch, *The Railway Journey: The Industrialization of Time and Space in the Nineteenth Century* (Berkeley: University of California Press, 1986), pp. 188–97. See also the "Oz" novelist and shop decorator L. Frank Baum's comments on the effects of mechanically moved displays in shop windows: "People are naturally curious. They will always stop to examine any thing that moves and will enjoy studying out the mechanism or wondering how the effect has been maintained." (Baum, *The Art of Decorating Dry Goods Windows* [Chicago, 1900], p. 87, as cited by Stuart Culver, "What Manikins Want," *Representations* 21 [Winter 1988], 107.) Such a "floating" of the difference between persons and things—a difference located in the principle of mobility or locomotion and floated (as we will see in a moment) by the uncanny juxtaposition of still bodies and moving machines—makes for the fascination of the still life of the market (that is, the fascination with the "mechanism" or "effect" of personation through illusion) in instances such as these.

45. See Sohn-Rethel, *Intellectual and Manual Labor*, p. 121.
46. Karl Marx, *The Grundrisse* (New York: Harper and Row, 1971), p. 132.
47. Karl Marx, *Capital*, vol. 1 (Harmondsworth: Penguin Books, 1976–81), pp. 301–05.
48. Thorstein Veblen, *The Place of Science in Modern Civilization and Other Essays* (New York: B.W. Huebsch, 1919), pp. 29–31, 93, 145.
49. Daniel Bell, *The Cultural Contradictions of Capitalism* (New York: Basic Books, 1976), p. xxv.

Part V: The Love-Master

1. Ernest Thompson Seton, *Boy Scouts of America: A Handbook of Woodcraft, Scouting, and Life-craft* (New York: Doubleday, Page, 1910), pp. xi, xii, 1–4, 34–38; Seton, "History of the Boy Scouts," p. 10, as cited by Michael Rosenthal, *The Character Factory: Baden-Powell and the Origins of the Boy Scout Movement* (New York: Pantheon, 1986), p. 65. On the relations and rivalries between the Woodcraft and scouting movements, and between Seton and Baden-Powell, see, in addition to Rosenthal, David I. Macleod, *Building Character in the American Boy: The Boy Scouts, YMCA, and Their Forerunners, 1870–1920* (Madison: University of Wisconsin Press, 1983). Macleod, more generally, provides a lucid and informed account of the making of middle-class Americans at the turn of the century.
2. Theodore Roosevelt, "The Strenuous Life," in *The Works of Theodore Roosevelt*, Memorial Edition (New York: Scribner's 1924–26), 15: 267, 271.
3. Frederick Jackson Turner, *The Frontier in American History* (New York: H. Holt, 1920), pp. 1, 4. See also Roderick Nash, *Wilderness and the American Mind* (New Haven: Yale University Press, 1973), pp. 141–60.
4. Ronald T. Takaki, *Iron Cages: Race and Culture in Nineteenth-Century America* (New York: Alfred A. Knopf, 1979), pp. 253–79.
5. BSA, *Handbook for Scout Masters* (New York: National Council Boy Scouts of America, 1914), p. 102.
6. As Ernst Haeckel observed in his widely influential *The Riddle of the Universe* (New York: Harper, 1900): "We can only arrive at a correct knowledge of the structure and life of the social body, the state, through a scientific knowledge of the structure and life of the individuals who compose it, and the cells of which they are in turn composed" (p. 8). See also Stephen Jay Gould, *Ontogeny and Phylogeny* (Cambridge, MA:

Harvard University Press, 1978); and Leo W. Buss, *The Evolution of Individuality* (Princeton: Princeton University Press, 1987).

7. Seton correspondence, as cited by Macleod, *Building Character in the American Boy*, p. 101; G. Stanley Hall, *Adolescence: Its Psychology and Its Relations to Physiology, Anthropology, Sociology, Sex, Crime, Religion, and Education* (New York: D. Appleton, 1904), 2:648.
8. To anticipate: The notion of *wilding* governs at least the media accounts of the recent brutal attack on a woman by a "pack" of youths in that nature preserve at the heart of the city, the Central Park designed by Frederick Law Olmsted as a remedy for the "over-civilized" and degenerate urban dweller. The account invokes the naturalist idiom of a violent primitivism and regeneration: a regeneration that takes place in the reproduction of "the natural" represented by the park or "nature museum" and a primitivism, at once anti-natural and anti-female, that is here racialized, not merely in the races of the attackers and victim, but also in the racial idea of the "gang period" itself. Across from the Park stands the Museum of Natural History and Roosevelt Memorial, with its visual reproductions of Nature—nature as *nature morte*.
9. In addition to the sources already cited, see, for instance: H.W. Gibson, *Boyology: or Boy Analysis* (New York: Association, 1918); Thorton W. Burgess, "Making Men of Them," *Good Housekeeping* 59 (July 1914); William Byron Forbush, *The Boy Problem*, 6th ed. (Boston: Pilgrim Press, 1907).
10. Jean Baudrillard, *For a Critique of the Political Economy of the Sign*, trans. Charles Levin (St. Louis: Telos Press, 1981), p. 85.
11. Seton, *Boy Scouts of America: A Handbook*, pp. 3–4. The panic about "consumption" at the turn of the century makes reference at once to what is perceived as the national disease of "consumerism" and of course to the perceived spread of the more "literal" plague of consumption throughout the social body ("perceived" because the category of consumption loosely "covered" both a range of diseases and a range of mental dispositions). It's here the conflation of bodily, national, and economic states that most concerns me: for instance, Seton's racializing and gendering of "the white man's plague" of consumption in American culture.
12. *Congressional Record*, 47th Congress, 2d Sess., 14 (March 1, 1883), p. 3488.
13. Robert Baden-Powell, *Headquarters Gazette* 5 (November 1911), 2, as cited by Rosenthal, *The Character Factory*, p. 6.
14. See Macleod, *Building Character*, pp. xi, 28, 104–6. As the social historian Samuel Haber has shown, in his *Efficiency and Uplift: Scientific*

Management in the Progressive Era, 1890–1920 (Chicago and London: University of Chicago Press, 1964), "The literature of system leaned heavily upon analogies to the human body, the machine, and the military. The body and the machine usually illustrated the need for close integration within the factory while military organization exemplified hierarchy and discipline" (p. 19). The scouting organization efficiently coordinates the body and the machine, in its physical cultures, in its quasi-military troops of boys, and in its hierarchial organization.

15. Francis Galton, "Eugenics: Its Definition, Scope and Aim," *Nature* 70 (1904), 82; Karl Pearson, *Darwinism, Medical Progress, and Eugenics* (London: Dulan, 1912), p. 27; Arthur Conan Doyle, preface to *The Construction and Reconstruction of the Human Body* (London: John Bale and Davidson, 1907), p. x. On Sandow and the physical cultures movement generally, see Harvey Green, *Fit for America: Health, Fitness, Sport, and American Society* (New York: Pantheon Books, 1986).

16. Ernest Thompson Seton, *The Birch-bark Roll of the Woodcraft Indians* (New York: Doubleday, Page, 1906), p. 4. "Honors by standards"—or, more familiarly, the merit badge system—forms part of the more general merit system of standardized individualism I outline here.

17. The treatment of rival styles of individualism I have been setting out here and in the preceding parts of this study is indebted to Foucault's accounts of the "government of individualization" and the making of individuals and individuality as an effect of disciplinary technologies and systematic management. See, for instance, "The Subject and Power," *Critical Inquiry* 8 (Summer 1982) and *Discipline and Punish: The Birth of the Prison* (New York: Pantheon Books, 1977), esp. pp. 135–69. The following discussion also extends and redirects the account of individualization, discipline, and desire in the American culture of managerialism set out in my *Henry James and the Art of Power* (Ithaca: Cornell University Press, 1984), pp. 96–145.

18. James R. Beniger, *The Control Revolution: Technological and Economic Origins of the Information Society* (Cambridge, MA: Harvard University Press, 1986), pp. 49, 185; Alfred D. Chandler, Jr., *The Visible Hand: The Managerial Revolution in American Business* (Cambridge, MA: Harvard University Press, 1977), pp. 27, 1; Daniel Boorstin, *The Americans: The Democratic Experience* (New York: Random House, 1973), pt. 5. On "mechanical prime movers" and machine culture generally, see Thorstein Veblen's *The Place of Science in Modern Civilization and Other Essays* (New York: B.W. Huebsch, 1919).

19. For Veblen, the uneven transition from the invisible hand of the market to the visible one of managerialism is in fact seen as a tension between honors and standards—or, in Veblen's terms, between "exploit" and

"industry." The distinction between exploit and industry is, for Veblen, a distinction between the rival tendencies of market culture and machine culture. If the first remains attached to the body and its desires—to an "interpretation of human nature in terms of the market" (that is, in terms of "the sensation of consumption"), the second transcends body and desire both (*Place of Science*, pp. 141, 231–51; on Veblen, see also Parts I, II, and IV above).

Exploit, for Veblen, includes but is not limited to the "archaic survivals" of the predatory impulse evident in the militarism and "the boys' brigades and other quasi-military organizations," in the cult of the wild, and big game trophy-hunting that proliferated in the 1890s. These survivals are, in Veblen's view, not *alternatives* to market culture but rather *instances* of market culture itself. And these instances include not merely the "buccaneer" capitalism of corporate raids and head-hunting, the predatory "pecuniary exploits" of market culture and its "captains," but also the forms of competitive or possessive individualism and the rituals of competitive, or conspicuous, consumption. (Corporate head-hunting might form a subject of its own: in order to join Roosevelt's Boone and Crockett Club, it was necessary to have collected at least three trophy heads; Roosevelt himself had eight.) (See Veblen, *The Theory of the Leisure Class: An Economic Study of Institutions* [1899] [New York: New American Library, 1953], pp. 27–29, 170–74.)

Exploit remains, on all counts, bound up with the "radiant body": with the reassertions of physical prowess and with the understanding of the economy in "sensuous terms." The logic of exploit, which is also the logic of market culture, thus remains tied to "animistic and anthropomorphic explanations": to the understanding of the economy in terms of the "disintegrating animism" of an "unseen hand." "Industry"—what Veblen calls "the metaphysics of the machine technology"—involves, precisely, a transcendence of the natural body and such anthropomorphic and animistic explanations. The logic of industry calls for, in short, a transcendence of the "*ordre physique*." The logic of industry demands the replacement of the radiant body by the "disciplinary effects" of the "machine process" and by an *im*personal "body of matter-of-fact knowledge" (*Theory of the Leisure Class*, pp. 81–82, 28–31; *Place of Science*, pp. 1–31, 55, 82–113.)

The *commuting* between these rival logics—between the logic of market culture and possessive individualism, on the one side, and the logic of machine culture and disciplinary individualism, on the other—such a commuting might be taken to define the double logic of desire and discipline in the culture of consumption. (For an extended account of that double logic, with particular reference to the incitement of wants and standardization of interior states in realist-naturalist discourse, see my "Advertising America," in *Henry James and the Art of Power*, pp. 96–145.) But it may be noted that a good deal of recent work on turn-of-the-century American culture continues precisely the "interpretation of human nature in market terms"—the equations of economics and desire within the logic of possessive individualism and an abstractly, and anachronistically, conceived notion of "the market." It is not merely that the generalization or inflation of C.B. Macpherson's "political theory of possessive individualism" to explain later nineteenth-century social conditions makes for contradictions and historical inaccuracies. The "new historicist" account of the market is not historical but methodological or theoretical (see Parts II and IV). That account, we have seen, *posits* a tautological relation between individual desires and social demands—the logic of sheer culturalism. Finally, the persistence of the understanding of the visible hand of the managerial economy in terms of the animistic and invisible one of the market economy is, of course, one of the regenerative rituals of consumerism: consumption as a paying homage to the radiant body, albeit in mass-produced and standardized form. Consuming as the call of the wild. And if recent cultural criticism tends to ratify such an understanding, this is certainly not the only way in which recent criticism replays the panic about agency and self-possession that it takes as its subject.

20. Veblen, *The Theory of the Leisure Class*, pp. 26–29.
21. Henry Ford, *My Life and Work* (Garden City, NJ: Doubleday, Page, 1923), pp. 108–9.
22. James Howard Bridge, *The Inside History of the Carnegie Steel Company: A Romance of Millions* (New York: Aldine, 1903), p. 85.
23. Harry Braverman, *Labor and Monopoly Capital: The Degradation of Work in the Twentieth Century* (New York: Monthly Review Press, 1974), p. 125. See also: Howard P. Segal, *Technological Utopianism in American Culture* (Chicago: University of Chicago Press, 1985), pp. 19–32, 98–99. In at least one sense, then, the managerial historian and theorist Peter Drucker is exactly right in his euphoric assessment of scientific management: "Indeed, Scientific Management is all but a systematic philosophy of worker and work. Altogether it may well be the most powerful as well as the most lasting contribution America has made to Western thought since the Federalist Papers." (Peter F. Drucker, *The Practice of Management* [New York: Harper, 1954], p. 280.) Both systems of management and governance (the Federalist Papers and Taylor's instruction cards and the policy of "written instructions") coordinate representation, production, and control by way of the paper forms of the American "paper system." For a richly detailed account of forms of

scientific management and responses to it, see also Martha Banta, *Patterned Lives* (Chicago: University of Chicago Press, forthcoming).

24. Niklas Luhmann, "Modes of Communication and Society" (1984), repr. in *Essays on Self-Reference* (New York: Columbia University Press, 1990), pp. 99–106. See also Luhmann's "The Individuality of the Individual: Historical Meanings and Contemporary Problems" in *Reconstructing Individualism: Autonomy, Individuality, and the Self in Western Thought*, ed. Thomas C. Heller, et al. (Stanford: Stanford University Press, 1986), pp. 313–25 (also reprinted in *Essays on Self-Reference*).

25. We might consider here the episode recounted in Taylor's *Principles of Scientific Management*, the episode in which he in effect invents the cigarette break: Taylor's "task of systematizing the largest bicycle ball [ball-bearing] factory in this country." The work process in this factory is the work of examining and sorting out perfect from defective balls. Taylor's systematizing involves not merely the introduction of regular breaks, in line with the "personal coefficients" of the workers; he introduces a series of inspectors and "over-inspectors" who inspect and classify the work—that is, the work of inspection and classification—done by the young women. This potentially infinite regress in the work of looking and sorting instances the relocation of the work process as a process of representing, classifying, and information-processing. (See Frederick Winslow Taylor, *The Principles of Scientific Management* [1911; New York: Norton, 1967], pp. 86–97.) The resistance to the understanding of information-processing as "real work" has certainly not disappeared (it persists, for instance, in the notion of a move from an industrial to a post-industrial or "information society" or in the denigration of "paper-pushing"). See, for instance, the somewhat indecisive account of information-processing and the work process in Shoshana Zuboff's *In the Age of the Smart Machine: The Future of Work and Power* (New York: Basic Books, 1988), esp. pp. 58–123. Or consider this exchange between characters in Pynchon's *The Crying of Lot 49* (1966), which focuses on the explanation of the ostensible paradox about work in the case of Maxwell's demon: " 'Since the Demon only sat and sorted, you wouldn't have put any real work into the system. So you would be violating the Second Law of Thermodynamics, getting something for nothing, causing perpetual motion.' 'Sorting isn't work?' [the ex-secretary] Oedipa said." Thomas Pynchon, *The Crying of Lot 49* (New York: Bantam, 1967), p. 62.

26. Beniger, *The Control Revolution*, p. 10. On Loeb and biomechanics, see Philip Pauly, *Controlling Life: Jacques Loeb and the Engineering Ideal in Biology* (New York: Oxford University Press, 1987).

27. Cited by David F. Noble, *America by Design: Science, Technology, and the Rise of Corporate Capitalism* (New York: Oxford University Press, 1977), p. 321.

28. Muybridge's model is mentioned in Miles Orvell, *The Real Thing: Imitation and Authenticity in American Culture, 1880–1940* (Chapel Hill: University of North Carolina Press, 1989), p. 311.

29. Gilles Deleuze, *Masochism: Coldness and Cruelty* (New York: Zone Books, 1989), pp. 20, 69, 76.

30. References to *The Red Badge of Courage* are to the Norton Critical Edition, ed. Sculley Bradley, et al. (New York: Norton, 1976) and are included in parentheses in the text.

31. Klaus Theweleit, *Male Fantasies*, vol. 1, trans. Stephen Conway (Minneapolis: University of Minnesota Press, 1987), p. 233. What Crane's account makes visible is thus the identification of machine labor and mechanized battle: what Ernst Jünger called the "amazing identity of processes" in the conjoining of the "war front and the labor front . . . The soldier's uniform appears ever more clearly to be a special case of a labor uniform." Ernst Jünger, *Der Arbeiter: Herrschaft und Gestalt* (The Worker: Domination and Form) (1932), as cited by Jeffrey Herf, *Reactionary Modernism: Technology, Culture, and Politics in Weimar and the Third Reich* (Cambridge: Cambridge University Press, 1984), p. 105. Both the war front and the labor front involve, in the technocratic imaginary, the coordination of body and machine. As Anson Rabinbach summarizes it, "The human organism was considered a productive machine, stripped of all social and cultural relations and reduced to 'performance,' which could be measured in terms of energy and output." (Rabinbach, *The Human Motor: Energy, Fatigue, and the Origins of Modernity* [New York: Basic Books, 1990], p. 183.) Such a reduction to pure performance in turn makes possible what the historian Charles S. Maier has identified as the technocratic "effort to combine technology with vitalist sources of energy." Hence the emergence of a "militarism of production" that vitalizes technology and merges labor and war, such that war appears as "a gigantic labor process." (See Charles S. Maier, "Between Taylorism and Technocracy: European Ideologies and the Vision of Industrial Productivity in the 1920s," *The Journal of Contemporary History* 6:2 [1970], 27–63.) As I have noted in Part III: "Statistical Persons," Crane's account of the war machine provocatively anticipates the technocratic vision of the soldier-producers, engaged in the work of producing corpses: the soldiers (in Jünger's terms) as "the day laborers of death." (It is not at all surprising that Mussolini early on renamed his own newspaper the *Daily for Soldiers and Producers*.) More generally, Crane's account powerfully discloses the intimate links between the dis-

passionate sequences of cause and effect in technocratic thinking and the eroticized violence of the disciplined, collective, and regimented body of the corps: the merger of the vital body and moving machine in what amounts to "the operation of a turbine filled with blood" (Jünger, "Die totale Mobilmachung" [Total Mobilization], in *Werke*, Band 5, *Essays* I [Stuttgart, 1960–65], p. 120.)

32. Sigmund Freud, *New Introductory Lectures, The Standard Edition of the Complete Psychological Works of Sigmund Freud*, ed. James Strachey (London: Hogarth, 1953–74), 22:80.
33. Boorstin, *The Americans: The Democratic Experience*, p. 549.
34. Jack London, *White Fang*, in The Library of America edition of *Jack London: Novels and Stories* (New York, 1982), pp. 92, 161. References to *White Fang* (WF) and *The Sea-Wolf* (SW) are to this edition. References to *John Barleycorn* (JB) are to the Library of America edition of *Jack London: Novels and Social Writings* (New York, 1982). References to *The Call of the Wild* (CW) are to the casebook edition, ed. Earl J. Wilcox (Chicago: Nelson Hall, 1980). All references are included parenthetically in the text.
35. It would be possible to trace the links between the coldness and cruelty of this logic and what will reappear as the "white terror" of the fascist state. That is, it would be possible to explore, in "the pitiless, spectral syllogisms of the white logic" seen "from his calm-mad height" (*JB*, p. 940), the terroristic racial and sexual violence "naturalized" (in the sense I have tried to indicate) in London's stories of the great white male North. On the "White Terror," see Klaus Theweleit, *Male Fantasies*, vol. 2, trans. Erica Carter and Chris Turner (Minneapolis: University of Minnesota Press, 1989). American imperialism, at the turn of the century, is bound up with the syllogisms of such a white logic. Hence the set of analogies that structure Richard Harding Davis's observation that "the Central American citizen is no more fit for a republican form of government than he is for an arctic expedition." See *Three Gringos in Venezuela and Central America* (New York: Harper & Brothers, 1896), p. 146. For a highly informative account of the workings of the "icy will" and "steely romanticism" in fascist, but not only fascist, technocratic thinking, see Herf, *Reactionary Modernism*; I have in mind particularly the discussion of the technocrat's "magical" or transcendental realism: the opposition of a "permanent technology" to an "evanescent capitalism" in the question concerning technology (pp. 70–108). (On another connection between the white North and the white logic and "spectral syllogisms" of technophilia, see note 37, below.)
36. Robert Baden-Powell, *Young Knights of the Empire: Their Code and Further Scout Yarns* (London: C. Arthur Pearson, 1916), p. 163.

37. The abstract and emblematic character of the white spaces London's figures move across, and the lines and tracks traced across them (like the white winter landscape the train line crosses in Zola's *La Bête humaine*) might be taken to emblematize the practice of writing. But, as I have argued in the Introduction, these "lines" are to be seen less as writing in the "traditional" sense (for instance, in the sense of an extension or reflection of identity or self-identity) than as the pressured radicalization of writing in general made possible by new technologies of inscription, impression, and registration. That is, these lines and tracks traced against abstract backgrounds appear as *une langue inconnue* of the body-machine writing itself (see Introduction). They appear as part of the transformation in the understanding of processes of representation, inscription, and information *as* production: the identification of processes of representation or inscription and physical processes I have been setting out in this chapter.

I have been arguing that the body-machine complex is inseparable from the naturalist mechanics of writing. The work of the body-machine writing itself, as it moves across and inscribes "its toil of trace and trail" across blank spaces, appears again and again in London's writing. London's mechanics of writing are everywhere bound up with the identification of the life process and the machine process: with the understanding of life in terms of the mechanics of bodies in motion and in terms of the inscriptions traced by these motions. Consider, for instance, this description of the main character at the close of London's story, "Love of Life":

> There were some members of a scientific expedition on the whale-ship *Bedford*. From the deck they remarked a strange object on the shore. It was moving down the beach toward the water. They were unable to classify it, and, being scientific men, they climbed into the whale-boat alongside and went ashore to see. And they saw something that was alive but which could hardly be called a man. It was blind, unconscious. It squirmed along the ground like some monstrous worm. Most of its efforts were ineffectual, but it was persistent, and it writhed and twisted and went ahead perhaps a score of feet an hour. ("Love of Life," in *Jack London: Novels and Stories*, p. 436; originally published in *MacClure's Magazine* 26 [December 1905])

This is the decomposition of the life process into systems of classification and the abstraction of moving natural bodies ("it writhed and twisted") into measurements of directed and effectual or ineffectual effort: the scientific measurement and registration of time and motion ("a score of feet an hour").

London's writings, particularly his accounts of the zero-degree white North, frequently take the mathematical form of a count-down or calibrated dissipation of energy within a closed system. For example, in the well-known story "To Build a Fire" (1908) or in the story of terminal cabin fever, "In a Far Country" (1899) or in the opening sequence of *White Fang*: in such stories, heat, energy, and the capacity for motion are depleted one by one by one as the system—the natural body or its technological prostheses—approaches degree-zero or entropy. The plot of such stories might be plotted on a graph or reduced to the calculations "n–1, n–2, n–3. . . . n–n." The internal relations between writing/calculating and the body-machine complex could not be more clearly marked.

In sum, the materialist *reduction* of the life process to matter and mechanics is also the systematic *abstraction* of the life process. The reduction of persons to bodies and bodies to sheer matter is the abstraction of persons, bodies, and matter to systems of representation, inscription, calculation, and measurement. This is the double discourse of transcendental materialism and the naturalist machine. And this is what makes it possible for technologies of writing, registration, calculation, and information-processing, on the one side, and the motions of the body-machine complex, on the other, to communicate at every point.

38. See also, CW: "His muscles became hard as iron . . . He could eat anything, no matter how loathesome or indigestible, and, once eaten, the juices of his stomach extracted the last particle of nutriment . . . building it into the toughest and stoutest of tissues" (p. 25).

39. See also, WF: "Another advantage he possessed was that of correctly judging time and distance. Not that he did this consciously, however. He did not calculate such things. It was all automatic. His eyes saw correctly, and the nerves carried the vision correctly to his brain. The parts of him were better adjusted than those of the average dog. They worked together more smoothly and steadily . . . When his eyes conveyed to his brain the moving image of an action, his brain, without conscious effort, knew the space that limited that action and the time required for its completion" (p. 203). London's consummate version of the call of the wild as the call to systematic management also invokes the later nineteenth-century updating of the Cartesian notion of the animal-machine. See William Coleman, *Biology in the Nineteenth Century: Problems of Form, Function, and Transformation* (Cambridge: Cambridge University Press, 1977), pp. 120–30.

40. On the discipline of desire in the naturalist workplace, with particular reference to the erotic violence of segregation and transgression (walls

and openings) in Zola's fiction, see my *Henry James and the Art of Power*, pp. 181–83.

41. On what I am here calling the compulsory perversity of "modern sexuality," see Jean Laplanche, *Life and Death in Psychoanalysis*, trans. Jeffrey Mehlman (Baltimore: The Johns Hopkins University Press, 1976); and Susan Stewart, "The Marquis de Meese," *Critical Inquiry* 15 (Autumn 1988), 162–92.

42. Oliver Wendell Holmes, "The Stereoscope and the Stereograph," *The Atlantic Monthly* 3 (June 1859), repr. in Beaumont Newhall, ed., *Photography: Essays and Images* (New York: Museum of Modern Art, 1980), p. 60. In his "The New Story-Tellers and the Doom of Realism" (1894), William Thayer also makes explicit the association of realism and skin games—with what he calls "epidermism." See *Realism and Romanticism in Fiction: An Approach to the Novel*, ed. Eugene Current-Garclia (Chicago: Scott, Foresman, 1962), p. 158.

43. I am here indebted to Donna Haraway's extraordinary essay, "Teddy Bear Patriarchy: Taxidermy in the Garden of Eden, New York City, 1908–36," repr. in *Primate Visions: Gender, Race, and Nature in the World of Modern Science* (New York: Routledge, 1989), pp. 26–58.

44. Ernest Hemingway, *The Sun Also Rises* (New York: Charles Scribner's Sons, 1926), p. 72.

45. Antonin Artaud, *Van Gogh, the Man Suicided by Society*, trans. Mary Beach and Lawrence Ferlinghetti, in *Artaud Anthology* (San Francisco: City Lights Books, 1965), p. 158.

INDEX

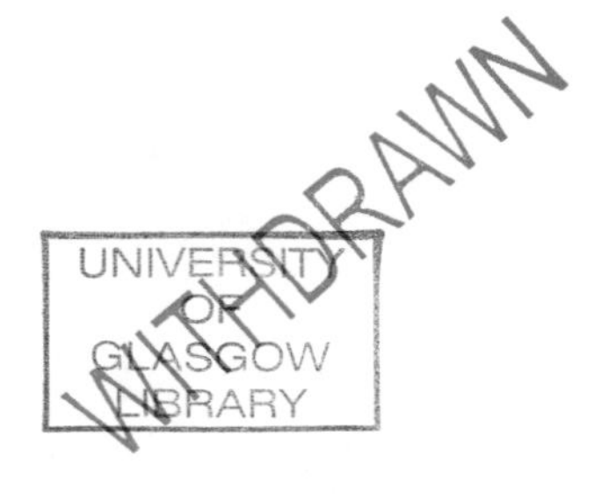